THE RELATIVISTIC DEDUCTION

BOSTON STUDIES IN THE PHILOSOPHY OF SCIENCE

EDITED BY ROBERT S. COHEN AND MARX W. WARTOFSKY

VOLUME 83

M. ÉMILE MEYERSON
(dessin de A. Bilis)

ÉMILE MEYERSON

THE RELATIVISTIC DEDUCTION

Epistemological Implications of the Theory of Relativity

Translated from the French by
David A. and Mary-Alice Sipfle

With a Review by Albert Einstein
and an Introduction by Milič Čapek

D. REIDEL PUBLISHING COMPANY

A MEMBER OF THE KLUWER ACADEMIC PUBLISHERS GROUP

DORDRECHT / BOSTON / LANCASTER

Library of Congress Cataloging in Publication Data

Meyerson, Emile, 1859–1933.
    The relativistic deduction.

    (Boston studies in the philosophy of science ; v. 83)
    Translation of: La déduction relativiste.
    Bibliography: p.
    Includes index.
    1.   Relativity (Physics)   2.   Science–Philosophy
I.   Title.   II.   Series.
Q174.B67   vol. 83   501 s   [530.1'1]   84–18152
[QC173.585]
ISBN-13: 978-94-010-8805-3   e-ISBN-13: 978-94-009-5211-9
DOI: 10.1007/978-94-009-5211-9

Published by D. Reidel Publishing Company,
P.O. Box 17, 3300 AA Dordrecht, Holland.

Sold and distributed in the U.S.A. and Canada
by Kluwer Academic Publishers
190 Old Derby Street, Hingham, MA 02043, U.S.A.

In all other countries, sold and distributed
by Kluwer Academic Publishers Group,
P.O. Box 322, 3300 AH Dordrecht, Holland.

This book was originally published by Les Editions Payot
in 1925 under the title *La Déduction Relativiste*.

TABLE OF CONTENTS

1. The Role of Mathematics. — 2. How Positivism Explains this Role. — 3. The Inadequacy of this Explanation. — 4. The Importance of Quality. — 5. Quality and Action. — 6. Quantity and the Nature of Things. — 7. Change and its Explanation. — 8. The Artistic Point of View. — 9. Conflicts and their Resolution. — 10. The Flux of the Quantitative. — 11. The Intelligibility of Reality. — 12. Auguste Comte's Protest.

13. The Preservation of Reality. — 14. Sensation and the Object. — 15. The Search for Consistency. — 16. The Objects Created by Science. — 17. The Attitude of the Philologist. — 18. Reality and Appearance. — 19. The Positivistic Point of View. — 20. Transcendence. — 21. The True Place of Theory in Science. — 22. Planck on the Retreat from 'Anthropomorphism'.

23. The Agreement between Mathematics and Reality. — 24. The Quantitative in Space. — 25. Deduction According to Descartes and According

EDITORIAL PREFACE

When the author of *Identity and Reality* accepted Langevin's suggestion that Meyerson "identify the thought processes" of Einstein's relativity theory, he turned from his assured perspective as historian of the sciences to the risky bias of contemporary philosophical critic. But Émile Meyerson, the epistemologist as historian, could not find a more rigorous test of his conclusions from historical learning than the interpretation of Einstein's work, unless perhaps he were to turn from the classical revolution of Einstein's relativity to the non-classical quantum theory. Meyerson captures our sympathy in all his writings: " . . . the role of the epistemologist is . . . in following the development of science" (250); the study of the evolution of reason leads us to see that "man does not experience himself reasoning . . . which is carried on unconsciously," and as the summation of his empirical studies of the works and practices of scientists, "reason . . . behaves in an altogether predictable way: . . . first by making the consequent equivalent to the antecedent, and then by actually denying all diversity in space" (202). If logic – and to Meyerson the epistemologist *is* logician – is to understand reason, then "logic proceeds *a posteriori*." And so we are faced with an empirically based Parmenides, and, as we shall see, with an ineliminable 'irrational' within science. Meyerson's story, written in 1924, is still exciting, 60 years later.

For Professor Čapek's illuminating introductory essay on this aspect as well as on the major themes of Meyerson's work and its context we are most grateful. We are grateful also for the splendid translation provided by the arduous and sensitive efforts of David and Mary-Alice Sipfle. We acknowledge also with thanks the kind cooperation of the Estate of Albert Einstein for permission to use Einstein's writings in this edition; and especially for Einstein's critical review of Meyerson's book; the help of Peter Mclaughlin with the German texts; the courtesy of John Stachel in providing material from the Einstein Project archive of the Princeton University Press.

Center for Philosophy and History of Science   ROBERT S. COHEN
Boston University

Department of Philosophy,   MARX W. WARTOFSKY
Baruch College, City University of New York

July 1984

TRANSLATORS' ACKNOWLEDGMENTS

Preparation of this volume was made possible by grants from the Bush Foundation, the translations program of the U.S. National Endowment for the Humanities, the *Direction du Livre* of the French Ministry of Culture and the *Maison des Sciences de l'Homme*, and the Boston Philosophy of Science Association. We especially want to thank Dean Peter Stanley and the members of the Carleton College Bush Grant Committee, Dr. Susan Mango of the National Endowment, Clemens Heller of the *Maison des Sciences de l'Homme*, and Robert S. Cohen of Boston University and the Boston Philosophy of Science Association for their help and encouragement.

We also want to thank our daughter, Gail Sipfle, our good friend David Porter of the University of Massachusetts, and our many friends and colleagues at Carleton College who gave freely of their time and wisdom and offered valuable advice when we shared translation problems with them: Sandra Allen, Debra DeBruin, Jackson Bryce, Linda Clader, Beverlee DeCoux, Sandra DuChene, John Dyer-Bennet, Roy Elveton, James Finholt, Evelyn Flom, Nancy Gustafson, Johanne Hall, Carl Henry, Andrea Iseminger, Gary Iseminger, Russell Langworthy, Perry Mason, Richard Miller, Thomas Moore, Bruce Morton, Richard Noer, Diethelm Prowe, Kim Rodner, Donald Schier, Susan Schlaeger, Daniel Sullivan, Carl Weiner, and Frank Wolf.

Most important of all, we wish to thank Milič Čapek not only for his suggestion that we undertake this translation, for his help and encouragement and for his fine introduction, but for all that we have learned from him in the years we have known him as teacher, colleague and friend. We dedicate this translation to him.

BIBLIOGRAPHIC ABBREVIATIONS AND
TRANSLATORS' COMMENTS

The following frequently cited works are referred to in the text or the notes only by an abbreviated title, as indicated below:

WORKS BY ÉMILE MEYERSON

ES: *De l'explication dans les sciences* (Paris: Payot, 1921).

IR: *Identité et réalité*, 2nd ed. (Paris: Alcan, 1912). Meyerson's citations are followed by pagination for the English translation, *Identity and Reality*, trans. Kate Loewenberg (New York: Macmillan, 1930 (Dover reprint, v.d.)). This is a translation of the 3rd ed. (1926) of *Identité et réalité*.

WORKS BY ALBERT EINSTEIN

ER: *Ether and the Theory of Relativity*, trans. G. B. Jeffery and W. Perrett, in *Sidelights on Relativity* (London: Methuen, 1922). This is a translation of *Äther und Relativitätstheorie* (Berlin: Springer, 1920). Meyerson cites *L'éther et la théorie de la relativité*, trans. Solovine (Paris: Gauthier, 1921).

GE: *Geometry and Experience*, trans. G. B. Jeffery and W. Perrett, in *Sidelights on Relativity*. This is a translation of *Geometrie und Erfahrung* (Berlin: Springer, 1921). Meyerson cites *La géométrie et l'expérience*, trans. Solovine (Paris: Gauthier, 1921).

RSG: *Relativity, the Special and General Theory* (New York: Crown, 1961). This is a translation of *Über die spezielle und die allgemeine Relativitätstheorie*, 10th ed. (Brunswick: Vieweg, 1920). Meyerson cites *La théorie de la relativité restreinte et généralisée*, trans. J. Rouvière (Paris: Gauthier, 1921).

VVR: *Vier Vorlesungen über Relativitätstheorie* (Brunswick: Vieweg, 1922). Meyerson's citations are followed by pagination for the English translation: *The Meaning of Relativity*, trans. Edwin Plimpton Adams, 2nd ed. (Princeton: Princeton University Press, 1946).

WORKS BY OTHERS

Borel (ET): Emile Borel, *L'espace et le temps* (Paris, 1922). Meyerson's citations are followed by pagination for the English translation: *Space and Time*, trans. Angelo S. Rappoport and John Dougall (New York: Dover, 1960).

Brunschvicg (EH): Léon Brunschvicg, *L'expérience humaine et la causalité physique* (Paris, 1922).

Cassirer (ER): Ernst Cassirer, *Zur Einstein'schen Relativitätstheorie* (Berlin, 1921). Meyerson's citations are followed by pagination for the English translation: *Einstein's Theory of Relativity* (bound with, and preceded by, his *Substance and Function*), trans. William Curtis Swabey and Marie Collins Swabey (Chicago: Open Court, 1923; reprint New York: Dover, 1953).

Comte (CPP): Auguste Comte, *Cours de philosophie positive*, 4th ed. (Paris, 1887).

Eddington (STG): Sir Arthur Eddington, *Space, Time and Gravitation* (Cambridge: Cambridge University Press, 1920; reprint New York: Harper Torchbook, 1959). Meyerson cites *Espace, temps, gravitation*, trans. J. Rossignol (Paris, 1921).

Langevin (AG): Paul Langevin, 'L'aspect général de la théorie de la relativité,' *Bulletin scientifique des étudiants de Paris*, no. 2 (April–May, 1922), pp. 2–22.

Langevin (PR): Paul Langevin, *Le principe de relativité* (Paris, 1922).

Petzoldt (SR): Joseph Petzoldt, *Die Stellung der Relativitätstheorie in der geistigen Entwicklung der Menschheit* (Dresden, 1921).

Planck (PRB): Max Planck, *Physikalische Rundblicke* (Leipzig, 1922). Meyerson's citations are followed by pagination for the English translation: *A Survey of Physical Theory*, trans. R. Jones and D. H. Williams (New York: Dover, 1960), an unaltered republication of *A Survey of Physics* (London, 1925).

Weyl (STM): Hermann Weyl, *Space – Time – Matter*, trans. Henry L. Brose (London: Methuen, 1921; reprint New York: Dover, 1952). This is a translation of *Raum, Zeit, Materie*, 4th ed. (Berlin, 1921). If a German edition is not indicated in our citations, Meyerson is quoting from the French translation of the same edition: *Temps, espace, matière*, trans. G. Juvet and R. Leroy (Paris, 1922).

Wien (WW): Wilhelm Wien, *Aus der Welt der Wissenschaft* (Leipzig, 1921).

Other works, after an initial full citation in each chapter, will be referred to in the text or the notes by name or short title only.

COMMENTS ON DOCUMENTATION

For works written in English we will normally cite only the English edition, even if Meyerson is working from a French translation. Where Meyerson is using French translations of non-English works, we will refer only to standard English translations from the original language when such translations are available. Where Meyerson is working directly from the original language, we will identify both the work he cites and the translation we are using. If we mention no published English language version, we are translating directly from the text provided by Meyerson.

Although Meyerson was a prodigious reader and obviously cared about his sources, there are a surprising number of errors in his quotations and his documentation. One has the impression he worked from cold, handwritten notes — or possibly, in some cases, from memory — and that he did not recheck his sources. Since reproducing careless errors seems to us to serve no scholarly purpose, we have corrected quotations freely and without comment unless there is a significant difference between Meyerson's version and the original, in which case we have so noted. We have freely added bibliographic information not provided by Meyerson himself and have silently corrected obvious errors where possible.

In a few cases we have been unable to locate Meyerson's quotations on the basis of his documentation or could not identify a source unambiguously. In these cases we have simply reproduced his notes and/or noted obvious errors in brackets. Although we have often corrected or completed bibliographic material without introducing brackets, all unbracketed explanatory notes are Meyerson's. Where Meyerson himself has supplied bracketed material, we have so indicated. Bracketed material not otherwise designated has been added by the translators.

INTRODUCTION

When Émile Meyerson's book *La déduction relativiste* was published in 1925, Einstein, in concluding his extensive review of it, called it "one of the most valuable contributions on the theory of relativity which has been written from the viewpoint of epistemology." This itself would be a sufficient reason for making it accessible to English and American readers, even though it appeared more than a half century ago. Despite the fact that a number of valuable philosophical commentaries on relativity theory have been published since then, Meyerson's book will always remain an important document concerning the early interactions between philosophy and the new physics. But the book itself is more than a mere historical document: What Einstein called its "unique" character is due to the author himself, whose background and attitudes were if not unique, then certainly exceptional. No easy and simple formula or label will do justice to the richness and complexity of Meyerson's thought. He was clearly a philosopher, more specifically a philosopher of science and epistemologist; but at the same time he was thoroughly acquainted with both chemistry and physics, especially of the classical period. Unlike most philosophers of science his knowledge was not confined to the content of the exact sciences; he was also very well informed about the origin and development of scientific concepts. This provided him with a wider perspective which freed him from what A. N. Whitehead called "provincialism in time," especially when his solid knowledge of the history of philosophy enabled him to see the intellectual currents and interactions between science and philosophy which minds confined to the narrow present period of contemporary science often fail to perceive. Thus the least misleading characterization of Meyerson is probably 'historian of ideas' — of the ideas in exact sciences and in philosophy.

The range of Meyerson's interests is so wide that it may easily arouse the suspicion of dilettantism, especially in the age of compartmentalization of knowledge. In the words of Ortega y Gasset, the average contemporary scientist "is only acquainted with one science and even of that he only knows the small corner in which he is an active investigator. He even proclaims it as a virtue that he takes no cognizance of what lies outside the narrow territory specially cultivated by himself, and gives the name of 'dilettantism' to any curiosity for the general scheme of knowledge."[1] Nothing would be more unfair than to charge Meyerson with dilettantism. One may occasionally disagree with his views, but none of them is formulated carelessly or hastily, each being supported by lavish documentation. In truth, his passion for documentation is almost obsessive when he conscientiously quotes verbatim whole passages of present and past authors to convince the reader that he does not distort or misinterpret their views. This leads him to certain verbosities and repetitions which led one early American critic[2] to say that "it takes a certain degree of patience to read it and in these busy times this is perhaps a reason why he has not been more influential." Since the times are even more busy now it is understandable why only one of his books — *Identité et réalité* (1907) has been translated in English so far. In this respect the present excellent translation of David and Mary-Alice Sipfle fills at least in part a gap which should have been filled long ago.

It is true that *Identity and Reality* contains all the central theses of Meyerson's philosophy of science and that the books which he subsequently published — *De l'explication dans les sciences* (1921), *La déduction relativiste* (1925), *Du cheminement de la pensée* (1931), *Le réel et le déterminisme dans la physique quantique* (1933) and even his posthumous *Essais* (1936) consist mainly in upholding, defending and documenting what he formulated in 1907. Obviously, there were new discoveries in both physics and chemistry, but many of them — although not all — easily yielded to Meyerson's interpretations. There was understandably no mentioning of relativity theory in the first edition of *Identity and Reality*; several footnotes referring to it were added to the third edition. But there is more discussion of it in his second book, while *La déduction relativiste* is almost exclusively focussed on it. This discussion can be adequately evaluated only when we recall the central themes of Meyerson's epistemology.

It consists in the view that the fundamental function of reason is to *understand* the phenomena; to understand means to *explain*; finally, any satisfactory explanation consists in *eliminating diversity, both in space and in time* by discovering the underlying *identities*. Briefly, according to Meyerson, *to explain means to identify*.

Stated in such a general way Meyerson's view appears paradoxical and even counterempirical, since experience is incurably diverse; diversity and change (which is nothing but diversity in time) is *given* in both sensory and introspective experience. This Meyerson does not deny; he even stresses that this diversified and changing given *cannot* be eliminated, even though our reason *imposes* on it its own identifying schemes as much as possible. This he shows in a very concrete, detailed and documented way and no brief summary can replace the wealth of his documentation. He shows how the *search for identity in time* is the most characteristic feature of both science and philosophy (whose separation took place only relatively recently, in the post-Kantian era). It appeared at the very dawn of Western thought, with the search of the Presocratics for the first principle, for the single unchanging ground, underlying the apparent diversity of the world as we perceive it. This monistic tendency found its most extreme formulation very early – in Eleatic philosophy. Very few thinkers followed Parmenides' 'acosmism' (this is Meyerson's term for the radical denial of both diversity and change, i.e., their complete dissolution in the undifferentiated unity of single and unchanging Being); so far as we know, only Zeno and Melissus accepted and defended his views.

But the fascination with the notion of a single timeless substance remained a lasting obsession of both science and philosophy. This can be clearly seen in the reaction of the Presocratic pluralists to Parmenides. Their reaction was a modification of the Eleatic principle rather than its repudiation. The atomists did not dare to deny the obvious reality of both diversity and change; but they did so with a minimum modification of the Eleatic principle. They preserved the complete homogeneity of Eleatic Being: all atoms are made of *the same stuff* as they differ only by their geometrical properties (position, shape and size), and their only change consists in changes in their relative positions. In other words, change is admitted only in its most innocuous sense – as a displacement – which leaves the qualitative unity of Being completely intact. It is true that Democritus called this Being "matter"; but his definition of matter remained Parmenidean – a mere impenetrable stuff, mere space-occupancy, completely devoid of any qualities and of any change. It is said that the atomists merely broke the solid Eleatic sphere into tiny bits and scattered them into the void; but each of its tiny bits retained all the characteristics – solidity, immutability and indivisibility – of Parmenidean Being. In admitting the existence of the void ('Non-Being' which the Eleatics denied) they made both diversity and change possible and thus made their system appear more empirical; but they were equally intransigent in their rejection of qualitative change and qualitative diversity.

With the coming of atomism it was also *mechanism* which entered the intellectual history of mankind, from which it has never really departed; its temporary eclipse in the Middle Ages was more due to the imposed authority of Christianized Aristotle than to any empirical discovery or argument. Even this temporary eclipse had been by no means complete, as Kurd Lasswitz, whom Meyerson quotes extensively in Ch. II, established convincingly. Its basic propositions were already stated above and in their more explicit form can be restated as follows: (a) matter is homogeneous and its quantity remains constant; it can be neither created, nor destroyed; (b) all qualitative differences in nature are merely apparent since they are due to the differences in configuration of the homogeneous basic units; (c) neither are the observed qualitative changes real since they result from changing configurations of the unchanging homogeneous units.

In this way the atomists anticipated with a remarkable accuracy *the kinetic-corpuscular models of nature* which were revived in the seventeenth century in a conscious opposition to Aristotle, and whose influence was considerably weakened only in this century. In particular, Meyerson pointed out that the principle of the conservation of matter was unambiguously stated by them more than two millenia before its experimental verification by Lavoisier. In this respect the atomists influenced even those pluralists who hesitated to accept their extreme monism, such as Empedocles and Anaxagoras; Empedocles, who retained the qualitative differences between the four elements, still insisted on their immutability: "of no one of all things is there any birth, nor any end in a baleful death. There is only a mingling and a separation of what is mingled." (Fr. 8.) Similarly Anaxagoras: "We Greeks are wrong in using the expression 'to come into being' and 'to be destroyed' for nothing comes into being or is destroyed. Rather, a thing is mixed with or separated from already existing things." (Fr. 17.) In other words, the qualitative pluralists, while upholding qualitative diversity, agreed with the atomists (and with Parmenides) in denying qualitative change since they reduced it to a mere *change of position*, i.e., to *motion in space*. As said before, the recognition that motion exists was a minimum concession made to experience.

But a concession it was, since a change of position is still a change and as such it resists the static schemes of the identification. It is then extremely interesting to follow Meyerson analyzing the devices by which the identifying intellect finally gets around this difficulty. It does so by formulating the *conservation laws*: to the law of the conservation of matter it adds that of the conservation of quantity of motion (momentum) and that of the conservation of energy. The exact formulation of both laws was obtained only in the modern post-Renaissance era and after considerable groping; Meyerson follows the development of the principle of inertia (which is the law of the conservation of momentum for a single body) in Chapter V and that of the law of the conservation of energy in Chapter VII. He pointed out that even today to unsophisticated 'common sense,' unacquainted with high school physics, the very existence of motion requires an explanation; motion cannot be self-sustaining, it requires an external *mover* as its cause. For this reason what may be called 'the law of the conservation of position' has for spontaneous human imagination a greater plausibility than the law of inertia: the material body not only requires an external mover to be forced out of its original position, but also needs it in order to be kept in motion. This was the view not only of Aristotle, but also of Kepler, who used the term *inertia materialis* for the alleged tendency of bodies to persist at rest and *return* to rest as soon as external force ceases to act upon them.

This persistent tendency to look for a mover behind every motion has not entirely disappeared even after the acceptance of the law of inertia and Newtonian mechanics in general; Meyerson mentions as a curiosity the fact that as late as in 1905 E. von Hartmann proposed at the meeting of *Société française de philosophie* that inertial motions result from the pressure of their cosmic environment! What is even more curious is that a similar hypothesis was favored half a century later by Viscount Samuel: inertial motion is due to the pressure of aether.[3] This urge to postulate a moving force behind every movement is merely a special instance of a more general tendency to *explain* the reality of change or rather *explain it away*. While changelessness seems to be rationally and logically self-sustaining, any change − including even a change of location − requires an explanation. Even the great Bernard Riemann, in other respects so free of the traditional prejudices, held the same view.

In this respect the atomists were far ahead of common sense and the Middle Ages since they come very close to the modern principle of inertia and the law of the conservation of momentum. Since in their original Leucippus-Democritean version they accepted not only the infinity, but also the isotropy of space, they had no difficulty in assuming that *the atoms move in all directions along straight lines* and change the direction of their motion only by clashing with one another. (Weight was not, as in the later Epicurean version, a primary property of atoms, but an effect of the surrounding whirling medium, that is, ultimately, of the impact of other atoms.) Clearly, both laws, that of inertia and of conservation of momentum were, if not explicitly anticipated, then at least foreshadowed by them; all that they lacked was adequate conceptual tools such as 'uniform velocity' and 'momentum,' which were forged much later. This anticipation Meyerson, usually so attentive to the intellectual ancestry of modern concepts, overlooked; it was explicitly recognized by Federigo Enriques and before him, by Hugo Carl Liepmann,[4] and implicitly by all those historians who were convinced that the space of Democritus, unlike the space of Epicurus and Lucretius, was isotropic.

But this minor oversight does not weaken Meyerson's main thesis about the way the intellect deals with motion: it *substantializes* it, i.e. it substitutes for it the concept of *the state* of motion or, more accurately, *quantity of motion* at a certain instant. In this sense, the author of *La déduction relativiste* speaks of 'substantialization of velocity' in the law of inertia. Since it would be contrary to our 'causal instinct' to accept that such a quantity can change without cause, its 'conservation' through time appears natural and the urge to find 'the identity in time' is satisfied. The atomists, in insisting that the motions exist from eternity and that their total amount remains constant, are obviously moving in the direction of the conservation laws. Evidently, the causal instinct '*a nihilo nit fit*' in its general form is too indeterminate and thus not very useful; if every change requires a cause, why not also change of position, as Aristotle and the scholastics believed? For this reason Meyerson rightly points out that the conservation laws cannot be *a priori* despite a number of ingenious attempts to prove it; the cooperation of experience is necessary to give any particular conservation law its concrete content. But it would be, according to him, equally misleading to speak of their exclusively *a posteriori* character as the inductivists believed; our experience is not entirely free since it is guided by the *causal tendency* to seek in the diversity of phenomena something which persists; and this tendency is, according to Meyerson *a priori*. (The term chosen by Meyerson to designate the character intermediate between *a priori* and *a posteriori* is 'plausible' – probably not a very good one.)

Although the conservation laws did recognize the existence of motion, its true nature was, semantically at least, rather obscured. Motion was regarded as a *substantial quantity, unchanging through time* while only its spatial distribution changes. This substantialization is even more conspicuous in the law of the *conservation of energy*. Meyerson and, two years later, the physicist Arthur Haas in his *Die Entwicklungsgeschichte des Satzes von der Erhaltung der Kraft* (Wien, 1909) traced in a thoroughly documented way its historical sources to the metaphysical and sometimes even theological principle of indestructibility and uncreatability of substance and its corollary – the equivalence of cause and effect. There is no place here to go over the detailed conceptual and historical analyses of both Meyerson and Haas. Suffice it to say that from the early rather qualitative statement of Lucretius in *De rerum natura* that no new force can suddenly appear in the universe and change the total amount of motion (*nequest in omne unde coorta queat nova*

vis irrumpere et omnem naturam rerum mutare et vertere motus[5]) up to the
seventeenth century, when Huyghens accurately formulated the law of conservation of energy for mechanics, and to the nineteenth century, when
Robert Mayer and Hermann von Helmholtz generalized it for the whole of
nature, there has been a very remarkable continuity of the same theme –
insistence on the uncreatability and indestructibility of motion. Apparent
exceptions were successfully explained by postulating invisible motions.
Leibniz was among the first to interpret the vanishing of the motion in the
collision of two inelastic bodies as a mere conversion of their visible kinetic
energy into the invisible kinetic energy of the corpuscles of which they both
consist; he thus foreshadowed the later notion of the mechanical equivalent
of heat. Similarly, the creation of the concept of potential energy explained
the cases when motion seemed to come '*ex nihilo*', as, for instance, when
the pendulum starts moving again downwards after reaching its uppermost
position where its kinetic energy was zero. Meyerson stressed that this new
concept of potential energy was interpreted by consistent mechanists in terms
of the kinetic energy of invisible particles; this was a natural consequence of
the basic assumption of mechanism that there is no action at a distance since
every interaction between bodies takes place only by direct contact. The
kinetic models of gravitation such as those proposed by Huyghens (and later
by Le Sage, etc.) were not excluded even by Newton. This is what Meyerson
called "le cinétisme pur" and its persistence may be traced to the first decade
of this century. On this point he could have easily increased his documentation by referring to C. Isenkrahe's survey of the kinetic theories of gravitation
(*Das Rätsel der Schwere*, Braunschweig, 1879) and to a later, more inclusive
survey by Paul Drude ('Über Fernwirkungen', *Annalen der Physik* 62, 1897,
pp. XIX–XLIX), or Osborne Reynolds' *The Submechanics of the Universe*
(Cambridge Univ. Press, 1903).

Meyerson and, after him, Arthur Haas had an easy task in showing how
important and even decisive were the aprioristic elements in the search for
the law of the conservation of energy. It was the causal principle, the equivalence of cause and effect, *causa aequat effectum*, which inspired both Robert
Mayer and Hermann v. Helmholtz and, long before them, Jean Bernoulli and
Leibniz: no quantity of energy, no matter how small, can disappear without
being transformed into an equivalent amount and, for the same reason, cannot appear *ex nihilo*, i.e., without being an equivalent of its antecedent form.
In other words, the law of the conservation of energy as well as that of matter
were, according to Meyerson, manifestations of the causal principle, of
identity in time. He could have quoted Kant, who regarded substance as an

a priori category whose quantity, according to the First Analogy of Experience, is constant through time; nothing better illustrates the persistence of this theme in Western thought than the fact that Kant quotes the words of the first century Roman poet Persius: *"Gigni de nihilo nihil, in nihilum nil posse reverti."*[6] Descartes' theological deduction of his law of the "conservation of motion" (which he failed to formulate correctly in a vectorial sense) is well known; Meyerson reminds us that two centuries later Joule's *a priori* conviction about the impossibility of the destruction of kinetic energy ($\frac{1}{2}mv^2$) remained unshaken by the rather disappointing results of his empirical verifications since, in his words, "it is manifestly absurd to suppose that the powers with which God has endowed matter can be destroyed any more than they can be created by man's agency." It is difficult not to recall in this context Einstein's words that "God does not play dice with the universe"; Meyerson would have certainly regarded his view as another manifestation of the causal principle, i.e., of the perennial tendency to postulate identity in time.

If the causal principle was frequently used to justify the law of conservation of energy, later the reverse process took place as well; both Herbert Spencer and Wilhelm Ostwald justified the principle *causa aequat effectum* by the law of the constancy of energy. The inequality of cause and effect would imply either annihilation or '*ex nihilo*' creation of a certain quantity of energy. It was thus understandable why the nineteenth century thinkers mentioned above regarded the law as having axiomatic certainty and that they, including Helmholtz, spoke of it with a metaphysical awe bordering on religious veneration. Hippolyte Taine, speaking of this law in his book *De l'intelligence* in 1870, exclaimed: "The immutable ground of Being has been attained; we had reached the permanent substance."[7] In another place, quoted by Meyerson in Appendix III of his second book *De l'explication dans les sciences*, the same author refers to "the supreme law" from which "as from its source" flow "by distinct and ramified channels the eternal torrent of events and the infinite sea of things." Similarly, to Ostwald energy is "the most universal substance" which inspired him to found his *Monistenbund* and his 'monistic Sunday sermons' (*monistische Sontagpredigten*). Herbert Spencer viewed this law as the most revealing manifestation of "the Unknowable Ultimate Reality" which he called also "Infinite and Eternal Energy," the respect for which would provide a common ground for a reconciliation of science and religion.

Nor did this almost religious respect for that law disappear with the coming of twentieth century physics. When the founder of wave mechanics, Louis de Broglie, in 1929 observed that since the only way to regard energy constant is to assign to it a definite value which is, however, excluded by the second form of the uncertainty principle, formulated two years before by Heisenberg, he concluded that the conservation law cannot be valid on the microphysical level ('Déterminisme et causalité dans la physique contemporaine,' *Revue de métaphysique et de morale* **38**, 1929, 442). Yet, it is significant that this article has never been reprinted in his subsequent writings, not even in those prior to his reconversion to classical determinism in 1952. This shows that he apparently always had some second thoughts about it, of which his return to classical determinism and to a deterministic interpretation of Heisenberg's uncertainty relations was, in a sense, a logical outcome. Meyerson, as already stated, never believed in the *a priori* character of the conservation laws in spite of repeated attempts to interpret them in such a way; the very fact that the discovery of the law of the conservation of energy emerged from the fruitless search for *perpetuum mobile* shows quite clearly that the uncreatability of energy had not always been as inconceivable as Herbert Spencer and Robert Mayer claimed.

Meyerson shows that besides the search for *identity in time* which in its most extreme form culminated in the *spatialization* and even elimination of time, there has been the parallel tendency to eliminate the qualitative diversity in space by the *geometrization of matter*. In its extreme form such *identity in space* was proposed together with identity in time only by Parmenides, but, as we shall see, some views of both philosophers and scientists came quite close to the *Eleatic acosmism*.

The unity of matter has always been the ideal of mechanistic explanation; the consistent atomists always insisted that there is *one* homogeneous matter underlying the observed qualitative differences. It is true that mechanism retained at least *one basic qualitative difference* – that between *empty* and *full* space, while the void has by definition only geometrical properties. The stuff which sporadically fills it, its *impenetrability*, remains, as Leibniz pointed out in his criticism of classical atomism, irreducible to geometry. Matter, of course, shares the properties of space, but, by the very fact that it *fills* or *occupies* space, it remains *irreducible* to it; in other words, the concept of matter implies that of space, but not *vice versa*. This distinction between the empty and the full has always been an insurmountable obstacle for a complete geometrization of matter.

There had been some other obstacles – Aristotle's theory of five qualitatively different elements and its residues in the modern period such as caloric and phlogiston; Dalton's qualitative atomism which posited the irreducibly different elements; the duality of mass and electricity. But none of these obstacles proved to be insurmountable: the theory of five elements collapsed when its underlying cosmological framework — the spherical geocentric universe — was removed; the differences among chemical elements were reduced to the differences in configuration of more basic particles — peripheral electrons and intranuclear particles, all of which, until the discovery of the neutron in 1932, possessed electrical charges, thus strengthening the belief in the fundamental unity of electricity and matter. Even the remaining duality of positive and negative electricity does not appear hopelessly irreducible, if we assume that the elementary charges instead of being indivisible, consist of the aggregations of the aether particles, as various hypotheses at the end of the last century suggested.

But not speaking of the enormous empirical difficulties which these hypotheses faced, admitting that the 'true atom' is the postulated 'aether particle,' would not the classical distinction of 'the full' and 'the void' still be present in the very same notion of the grain of aether? This is what was explicitly admitted by all granular models of the aether from Huyghens to their valiant defenders in the twentieth century. *Implicitly* this was conceded even by those who upheld various hydrodynamical models of the aether unless they would take its continuity and fluidity in a literal sense. This, needless to say, would be nothing but a thinly disguised return to the Aristotelian 'second element' — water — and thus would run contrary to the basic thesis of all corpuscular-kinetic models, which regard the apparent continuity of any fluid not as the primary physical quality, but as a mere *appearance*, hiding to our obtuse senses the lively motion of its constituent particles. Furthermore, Descartes, whom Meyerson regards as the modern initiator of the tendency to reduce the apparent physical diversity to modifications of the aether, was not, as Kurd Lasswitz pointed out, always consistent when he conceded very tiny separations between the particles of aether.[8]

To Descartes, Meyerson traces the concept of the all pervading continuous *plenum*; its historical roots go much farther, to Giordano Bruno's 'luminous air' and even to Stoicism. He correctly points out that the hydrodynamical models of Helmholtz, Larmor, Lord Kelvin and others are Cartesian in spirit. Certainly in such reduction of matter to the aether the tendency to spatialization of the world went as far as it could; Meyerson quotes Maxwell's view according to which the homogeneous, all-filling aether is indistinguishable from absolute space. But the difficulties of such a reduction of matter to space are insurmountable and most of them are listed by Meyerson. Even if we were to disregard the objections mentioned above against the very concept of the *plenum*, would any motion be possible in it? Descartes, in this respect following Aristotle, conceded that the only possible motion would be of a *vortical kind*, i.e., along a circle or any closed curve. Not speaking of the difficulty which such a notion implies, i.e., of instantaneous transmission of the pressure, Meyerson quotes another, even more serious objection, raised by Leibniz in a letter to des Bosses in 1706: that a rotational motion within a perfectly homogeneous fluid would produce no observable difference because the successive states of such a liquid cannot be distinguished from another state, "not even by an angel, and consequently there can exist no variety in phenomena." The same objection applies to the modern versions of the Cartesian models, such as William Thomson's 'vortex atoms,' and Meyerson recalls Poincaré's and Duhem's criticisms of such attempts. The greatest objection against the reduction of matter to space would be their success: the empirical world would simply vanish to be replaced by 'the sphere of Parmenides,' the self-identical substance, devoid of any differentiation and change.

But when the tendency to geometrize matter was not pushed to its absurd Eleatic extreme, its fruitfulness has been undeniable, and Meyerson shows its impressive successes in his books, first in the already mentioned Chapter VII, 'L'unité de la matière,' in *Identité et Réalité* and then in his second book *De l'Explication dans les Sciences*, in particular in Chapter VIII, 'Les modalités de l'explication spatiale.' He points out that whenever some qualitative diversity was encountered, there has always been the tendency to explain it in a Cartesian fashion *"par les figures et les mouvements."* He illustrates it by a number of examples from different periods; even some qualitative theories shared with the mechanistic theories the mode of explanation by *displacement* of certain quantities such as phlogiston or caloric.

When after Lavoisier and Dalton the qualitative diversity of the elements seemed to be definitely established, the hopes of finding the homogeneous substrate underlying their diversity had never entirely disappeared; what was merely a guess at the time of Prout, gradually has become — after Mendeleev, J. J. Thomson and Lorentz — a successful explanation of the intra-atomic structure. As Boyle rightly anticipated, the atoms of the elements proved not to be the ultimate units of matter since their diversity was a result of various configurations of electrons and protons which seemed to be the basic units of matter and electricity. But there was a rather disconcerting lack of symmetry in the relation of protons (positive electricity) and electrons (negative electricity) to mass; almost the whole mass of the atoms was due to the protons, since the mass of electrons was found to be almost two thousand times smaller. Meyerson, so far as I know, did not comment on this, although it would illustrate his thesis that any success of identification – in this case the identification of matter and electricity – brings the discovery of some new recalcitrant feature which spoils the desired unity; why are the protons so much *heavier* than the electrons?

The discovery of positive electrons of the same mass and charge as negative electrons restored a desired symmetry of two kinds of electricity, but only in part since it did not remove the striking inequality of the electronic and protonic mass. Also the discovery of the neutron ended, at least for a time, the hope to identify matter (mass) and electricity. These last two discoveries were made still during Meyerson's life time; he had no time to use them as new illustrations of the ongoing conflict between the identifying schemes and recalcitrant diversity of experience. His own term for this aspect of experience resisting identification was 'l'irrationnel.'

But the most striking examples of the success of 'spatial explanation' Meyerson found in chemistry, especially in the structural formulae of organic chemistry when an addition or subtraction of a single atom modifies the physical properties and chemical behavior of the molecule, or when a mere difference in the spatial arrangements such as in isomers produces similar changes. Examples abound, and Meyerson could have used them *ad libitum*. More recently, discoveries in molecular biology such as the discovery of the 'double helix' of DNA and of what Jacques Monod called the 'stereospecific properties' of proteins would have very much pleased Meyerson, who would have regarded them as new triumphs of explanation by configuration. The fact that a full understanding of the chemical link in organic compounds is impossible without considering the undulatory character of matter would have probably disturbed him – indeed, *did* disturb him – as his last book published in his lifetime, *Le réel et le déterminisme dans la physique quantique*, showed. On the other hand, the fact that he always regarded *sensation* or, more generally, *consciousness* as absolutely irreducible to spatial mechanistic models – the view which he shared with all epistemologically sophisticated thinkers of his time – should have prevented him from adopting the materialistic reductionism of the majority of Neo-Darwinians.

Was he consistent? In Chapter VII of his second book, 'Les phénomènes biologiques,' in which he surveyed the perennial conflict of mechanism and vitalism, he definitely preferred the former, and his doubts about the usefulness of the concept of 'entelechy', formulated by Driesch, proved to be rather prophetic. He concluded this chapter by the following very explicit statement: "In the whole immense field of science, there cannot be any real explanation except by space and by the properties of space." If it is so, should not the same explanation apply to sensation as well? This, however, Meyerson did not concede, as the irreducibility of quality to geometry was for him one of those irrational features which resist the identifying tendencies of the intellect.

The chapter of his second book mentioned above was one of a few places where Meyerson, whose background was mainly in physical sciences, in particular chemistry, discussed the problems of explanation in biology. Another passage of this kind is in the all-important Chapter V of the same book, 'Identité et identification,' dealing with the tendency *to eliminate time* in both science and philosophy. This tendency is closely correlated with the geometrization of nature; if the only change which is regarded as rational is *change of position*, which leaves intact the elements which are being displaced, the causal efficacy of time is eliminated. While this for the rank and file scientists did not imply the denial of time, those more sophisticated among them, such as Laplace, did not hesitate to deny its existence, or at least to reduce it to mere human ignorance. The equivalence of cause and effect, the conservation laws, the reversibility of the elementary mechanical processes, the reversibility of chemical equations – all these notions were, according to Meyerson, different manifestations of the same tendency. In the chapter, 'Identité et identification', Meyerson added to the examples from physical sciences some from the history of biology.

These examples are so much more striking because they come from the science dealing with growth, decay and what appears to be qualitative change. Yet, it was precisely qualitative change which various hypotheses of *preformation* tried to eliminate or – what is the same – to reduce to a change in size: the whole organism was supposed to preexist on a minute scale *in the ovum* (according to 'ovists') or in the sperm (according to 'spermists') and the only difference from its adult form was that of dimensions. This led logically to the bizarre hypothesis of the 'encapsulation of germs' (*l'emboîtement des germes*) according to which all successive human generations were contained either in the sperm of Adam or in the ovum of Eve! This led two eighteenth century Swiss biologists to the most extreme views; according to Charles Bonnet, "all pieces of the universe are contemporary; the Creative Will realized by a single act everything which could exist," while Albrecht von Haller simply declared that "there is no becoming." Later biologists were in this respect more careful, yet Meyerson points out – and he is not alone in his view – that a certain affinity of modern theories of heredity with the doctrine of preformation is undeniable. Jacque Loeb even did not hesitate to use the same term. From the views of Augustin Weissmann to the concept of the genetic code the overtones of preformationism are unmistakable, and Meyerson regards them as another evidence of "one eternal and immutable tendency of the human mind" which can be summed up in one sentence: "There is no novelty" (*Mais il n'y a pas de nouveau!*).[9] He recalls the ancient principle of atomism to explain all differences and changes in nature by "shape, arrangement and position," and he would have certainly welcomed Jean Rostand's book *L'atomisme en biologie*, which appeared more than two decades after Meyerson's death.

But the greatest empirical obstacle to a successful spatialization of time was for Meyerson the second law of thermodynamics, which in France is called 'Carnot's principle.' He deals with it in his three main books: in *Identity and Reality*, Ch. VIII, in *De l'explication dans les sciences*, Ch. VI and Appendix IV; finally in *La déduction relativiste*, Ch. VII. He points out an acute intellectual discomfort which the irreversibility of the processes of nature caused to both philosophers and scientists, and he analyzes various devices − some of them rather curious − invented in order to avoid the 'thermal death' of the universe which the inexorable increase of entropy implies. The basic motive of all such attempts was to extrapolate the principle of the reversibility of the elementary mechanical processes to the universe as a whole and thus to preserve, if not the strictly cyclical character of the cosmic history, then at least its general periodicity.

He traces this tendency to the very dawn of Western thought − to the Presocratics and the Stoics; the latter went even as far as to accept the doctrine of eternal recurrence, i.e., the periodical return of the same state of the universe in its most concrete details − the view revived in the modern period by Nietzsche and even as late as in 1927 by Abel Rey. In a less extreme form this tendency was present in the last century's cosmogonical theories of the 'from the nebula to nebula' type by which the rewinding of the cosmic clock would be assured; the speculations of Svante Arrhenius among others belong to this category as well as a curious hypothesis of W. Rankine about boundaries of the aether which act as the 'reflecting walls' of the universe, thus preventing the endless escape of radiation into outer space, and thus assuring the possibility of the reconcentration of energy. Meyerson could have pointed out a striking similarity of this hypothesis with the Stoic view for which aether was also a huge, although finite, bubble in limitless space.[10] Today he would regard the theory of the 'oscillating universe,' in which the periods of expansion alternate with those of contraction, as another illustration of the same tendency.

This tendency was to him another manifestation of the perennial search for 'identity in time,' i.e., ultimately of the elimination of time. He quotes the well known view of Ludwig Boltzmann, who considered the possibility that the entropy law may be only locally valid in our part of the universe while in some other very remote parts of space the entropy may be decreasing. This would mean, according to Boltzmann, that in the universe as a whole the 'forward' and 'backward' directions of time are as arbitrary and relative as the spatial directions 'up' and 'down.' Meyerson correctly saw that with the relativization of the 'direction' of time, succession itself loses its meaning and the very existence of time is in question. One can conceptually dissociate the orientation of the geometrical segment from the segment itself, but not succession from duration — unless one confuses time with its symbolic static representation. Thus it is hardly surprising that in Meyerson's personal encounter with Einstein he used 'the irreversibility of becoming', suggested by the second law of thermodynamics, as one of the decisive arguments against the spatialization of time.

This encounter took place on April 6, 1922 at a meeting of the French Philosophical Society (*Société française de philosophie*) at Paris. It was a rather unusual meeting since it was not a meeting of the French philosophical community only. Other outstanding members of French academic life were present: among physicists Paul Langevin, Jean Perrin and Paul Painlevé; among mathematicians Jacques Hadamard and Elie Cartan; among psychologists Henri Piéron. Conspicuous among philosophers were Henri Bergson, Émile Meyerson, Léon Brunschvicg and Edouard Le Roy. But attention was naturally focussed on the person of the very distinguished visitor — Albert Einstein. He was not a speaker in a usual sense; after being introduced by Xavier Léon and Paul Langevin, he was ready to answer questions. The question raised by Émile Meyerson consisted of two distinct parts, or rather of two distinct questions, both of them equally interesting.[11] One of them concerned Einstein's philosophical relation to Ernst Mach. Einstein's answer was significant as it corrected — or should have corrected — the widely spread view according to which he was a philosophical disciple of Mach. Einstein, while praising Mach as a physicist, characterized him at the same time as a 'poor philosopher' (*un philosophe déplorable*) mainly because Mach, blinded by his dogmatic phenomenalism, refused to concede the existence of atoms despite the overwhelming experimental evidence for it. (This probably did not do full justice to the last period of Mach's thought, when he conceded at least the usefulness of the atomistic hypothesis.) But a philosophically more interesting question was one which for Meyerson himself was more important since he asked it first — and Einstein also answered it first: is 'the spatialization of time,' i.e., the tendency to regard the relativistic time as 'the fourth dimension of space' a legitimate interpretation of Minkowski's concept of the union of space and time?

What prompted Meyerson to ask this question was the fact that the spatializing interpretation of relativistic time was quite frequent then and – we must add – still remains so, even today. Meyerson did not give any example of it for the simple reason that he would have been embarrassed to choose out of so many instances. Let me mention only two of them, both from the period prior to the year 1922. By coincidence they are both taken from books which appeared in 1914, i.e., only six years after Minkowski's historical paper:

With Minkowski, space and time become particular aspects of a single four-dimensional continuum; the distinction between them as separate modes correlating and ordering phenomena is lost, and the motion of a point in time is represented *as a stationary curve in four-dimensional space*. Now if all motional phenomena are looked at from this point of view, they become *timeless phenomena in four-dimensional space*. The whole history of a physical system is laid out as a *changeless* whole. (E. Cunningham, *The Principle of Relativity*, Cambridge, 1914, p. 191; italics mine.)

The second example is even more characteristic:

There is thus far an intrinsic similarity, a kind of coordinateness, between space and time, or as the Time Traveller, in a wonderful anticipation of Mr. Wells, puts it: "There is no difference between Time and Space except that our consciousness moves along it [to wit: along the former]. (L. Silberstein, *The Theory of Relativity*, London, 1914, p. 134)

This is explicit enough; all essential features of the static interpretation of Minkowski's world as it still lingers in the minds of some physicists and philosophers are present in these two passages.

Meyerson's objections to the static interpretation of Minkowski's concept were based on several arguments which were more systematically restated later in his new book *La déduction relativiste*. First, he pointed out that a certain asymmetry between space and time is still preserved in the relativistic union of both concepts: thus, while the space of Einstein is Riemannian, i.e., spherical, relativistic time does not have an analogous curvature. This is clear from the 'cylindrical model' of space-time which Einstein at that period accepted: the axis of this cylinder represented a linear irreversible time while the transversal circular cross-sections symbolized the finiteness and 'closed' character of its spatial instantaneous components. In Riemannian space it is in principle possible to return to the same point, but not to the same moment of time. He conceded that with the coming of relativity, something has changed in this respect because "time does not flow uniformly for everybody," as the traveller after a roundtrip made with a velocity approaching the velocity of light would find out by comparing his watch with that of the observer staying at home. Meyerson then added:

But there would be a limit to the divergence, for the traveler could never go backward in time. "One cannot telegraph into the past," Einstein quite rightly tells us, and out of the great upheaval which the theory of relativity necessitated for those concepts we believed the most firmly established, the principle of entropy is one of the two great principles of earlier physics to remain standing, the other being, as you know, the principle of least action. (p. 257 below)

Thus the rejection by Einstein himself of backward moving causal actions together with the law of entropy remaining intact in relativity represented Meyerson's second and third argument against spatialization.

His fourth argument was very probably inspired by Bergson:[12] the spatialization of time in certain philosophical interpretations of relativity is, according to him, merely the most recent manifestation of the perennial human tendency to treat time in a space-like fashion. (Surprisingly, Bergson, who attended the same meeting and also intervened in the discussion, did not raise this objection although the 'fallacy of spatialization' was the target of his criticism in all his writings, including the conclusion of *Durée et simultanéité*, published later the same year.

Meyerson's fifth and last argument should probably have been placed first since it underlies the rejection of backward moving actions. He recalled Weyl's proposal to speak of 'three plus one dimensions' instead of 'four dimensions' of space-time since in Minkowski's formula for the space-time interval the time variable is preceded by an algebraic sign opposite to those of the other three coordinates. Thus even the mathematical formalism of relativity theory reflects the heterogeneity of the temporal dimension.

What is surprising in Einstein's answer to Meyerson when we read it today is not so much his apparently complete agreement — Meyerson after all was quoting Einstein himself when he criticized the spatializing interpretations — but the briefness of his reply, which consisted in one single statement:

In the four-dimensional continuum, it is certain that all directions are not equivalent. (p. 260 below)

The rest of his answer dealt with his relation to Mach's philosophy, as mentioned above. Today in the light of Einstein's later utterances on this subject it is clear that his own attitude toward this problem was, as we are going to see, far from unambiguous.

This ambiguity was pointed out by Meyerson himself when he reformulated his objections against the static interpretation of space-time in his third book *La déduction relativiste* in 1925. While he recalled there again Einstein's words that "we cannot send telegraph messages into the past," he also quoted him saying that "the becoming in three-dimensional space is transformed into a being in the world of four dimensions." (See Ch. VII below, esp. pp. 71, 73.) The last statement could hardly have been more explicit; but, as we shall see, it was *not* his last word on this problem.[13]

For Einstein's response to Meyerson's criticism of 'the fallacy of spatialization' was enthusiastically positive. In a leading article which Einstein wrote for the *Revue philosophique* in spring 1928, Einstein highly praised Meyerson's book in the words quoted at the beginning of this *Introduction*. The very extent of this article certainly contrasted with the one sentence answer which Einstein gave to Meyerson six years before. His agreement with Meyerson's rejection of the spatializing interpretation of the world of Minkowski is evident from the following passage:

Thus he [Meyerson] rightly insists on the error of many expositions of relativity that refer to the 'spatialization of time.' Time and space are indeed fused in one and the same *continuum*, but this continuum is not isotropic. The element of spatial distance and the element of duration remain distinct in nature, even in the formula giving the square of the interval of two neighboring events. The tendency he denounces, though *often only latent in the mind of the physicist, is nonetheless real and profound, as is unequivocally shown by the extravagances of the popularizers, and even of many scientists, in their expositions of relativity.*[14]

It is thus evident that both Meyerson and Einstein at that time regarded the static interpretation of space-time as a grave error. The extent of Einstein's agreement with Meyerson can be also judged from the only reservation which he formulated about the book. For this reservation followed, paradoxically as it may sound, from his very agreement with Meyerson's thesis. Meyerson, like Bergson in his book *Durée et simultanéité* published several years before, regarded the general theory of relativity — as well as Weyl's and Eddington's attempts to incorporate gravitation and electricity into a single field theory — as continuations and even culminations of the old Cartesian tendency to 'geometrize matter,' i.e., to reduce all diversity of physical phenomena to geometrical differences. Einstein, while agreeing that such a tendency indeed exists and has always existed, said it can never fully succeed since a complete dissolution of matter into space remained a dream — the 'Cartesian dream,' as he called it — which neither classical physics nor the physics of relativity succeeded in realizing; concrete experience, in particular that of time, always resisted and resists it:

Consequently, I believe that the term 'geometrical' used in this context is entirely devoid of meaning. Furthermore, the analogy Meyerson sets forth between relativistic physics and geometry is much more profound. Examining the revolution caused by the new theories from the philosophical point of view, he sees in it the manifestation of a tendency already indicated by previous scientific progress, but even more visible here — a tendency to reduce "diversity" to its simplest expression, that is, to dissolve it into *space*. What Meyerson shows in the theory of relativity itself is that this complete reduction, which was the dream of Descartes, is in reality impossible. Thus he rightly insists on the error of many expositions of relativity . . . [*etc*. There follows the passage quoted above],

Two points should be stressed here. First, the reason why Descartes' original dream was condemned to a failure was that the only kind of space which he had at his disposal was the space of Euclid which, because of its complete homogeneity, cannot by any intellectual manipulation be identified with the concrete diversity of the physical content. (This is also true of two other spaces which in virtue of their constant curvature are equally homogeneous — the space of Riemann and that of Lobachevski.) Obviously, only spaces with their curvatures changing from one region to another could have the Cliffordian 'humps' which could simulate the presence of matter and its physical fields. But then the Cartesian ideal is, if not entirely given up, at least profoundly modified; true, the distinction between space and its physical content is eliminated, matter is truly 'resorbed' in space, as Meyerson says, but only at a price: its heterogeneity is really transferred to space itself in the form of its locally variable curvature which endows it with the causal efficacy which the Newton-Euclidean space did not have. In this way, as George Boas observed in 1931, "the drive toward an Eleatic unity seems to be arrested" as the elimination of the distinction between space and its physical content is obtained only by a heterogenization of space itself.[15] In Eddington's words, the so-called "geometrization" of mechanics could be equally well called "mechanization" or, as I would prefer to call it, "physicalization" or "dynamization" of geometry.[16]

This leads us to the second point: Meyerson's expression 'matter resorbed into space' (Ch. II below) is, if not incorrect, at least *elliptic* and thus misleading. For it is more correct to say that matter and its gravitational field which is inseparable from it is 'resorbed' not into space, but *space-time*; otherwise no motion, no change, would be possible. If this is often not explicitly stressed, it is because it is so obvious; without the presence of the activating 'temporal dimension' there would not be even classical kinematics, much less classical mechanics; this is so much more true in the physics of relativity where the union of space and time is so much closer. It is true that Meyerson's elliptical statement is nothing but a mere linguistic slip; but it is psychologically rather significant as it shows two features of our intellectual subconscious: first, the tendency to place space on an ontologically higher level than time, the tendency which Meyerson himself, following in this respect Bergson, so masterfully analyzed; second, the deeply ingrained traditional inclination, created by common sense and strengthened by Newtonian physics, to speak of space *separately* from time. (This shows itself even in modern treatises on cosmogony; we always hear of the expansion of *space*, rarely of *space-time*.) This only shows how much to the point Einstein's comment on Meyerson was when he stressed what Meyerson, according to him, did not emphasize enough: that it is misleading to speak not only of geometrization of time, but of geometrization of matter as well.

This, however, was not the last word of Einstein on this subject. When in 1949 Kurt Gödel wrote a vigorous defense of a static interpretation of relativistic space-time in a contribution to Schilpp's volume *Albert Einstein: Philosopher-Scientist* under the title 'A Remark About the Relationship between Relativity Theory and Idealistic Philosophy,' Einstein's comment on it was distinctly, though cautiously, sympathetic.[17] According to Gödel, the relativization of simultaneity destroys the objectivity of the time lapse and thus substantiates "the view of those philosophers who, like Parmenides, Kant and the modern idealists, deny the objectivity of change and consider change as an illusion or an appearance due to our special mode of perception." In a footnote he made a reference to the once-famous article of J. M. E. McTaggart, 'The Unreality of Time,' which appeared in *Mind* in 1908. As an additional argument for his view Gödel adduced the mathematical possibility of certain cosmological models in which one could "travel into any region of the past, present and future, and back again, exactly as it is possible in other worlds to travel to distant parts of space." Gödel even made an estimate of the quantity of fuel and the velocity of a space-ship to make such a fantastic trip. Such a rocket would be a realization of the 'time-machine' of H. G. Wells; Gödel would have subscribed to Silberstein's view regarding the famous British author as anticipating relativity. We would expect Einstein who so explicitly endorsed Meyerson's criticism of the static interpretation of Minkowski's world to reject resolutely such an extreme form of spatialization of time. But the very opposite took place. Did he forget his own explicitly stated view that "one cannot send telegraph messages to the past"? Is not Gödel's hypothetical rocket ship merely an oversized form of such a signal?

Indeed, Einstein apparently modified his view — perhaps under the influence of Hans Reichenbach's book which he recommended to Meyerson at the end of his comment — in the following way: the impossibility to send signals to the past holds only for macroscopic processes which are irreversible because of the steady increase of entropy; but it does not hold necessarily for the elementary microscopic processes which are allegedly reversible. Furthermore, if we concede with Gödel the possibility of closed world-lines on a huge megacosmic scale, the relation of succession itself is relativized since on a circular world-line it is a matter of convention to say that A precedes B rather than B precedes A. Similarly, Aristotle in considering the circularity of cosmic time claimed that we can with equal right claim that we live *prior* to the war of Troy as that we live *after* it.[18] It is true that Einstein added cautiously: "It will be interesting to weigh whether these [cosmological solutions] are not to be excluded on physical grounds." In truth, Gödel's universe would indeed be hospitable to causal anomalies of the weirdest kind of which even Nietzsche's cyclical universe would be free. For in the doctrine of 'eternal recurrence' advocated by Nietzsche and the ancient Stoics before him, *all* the world-lines would be closed, thus restoring identically the previous state of the universe in *all* its details, an odd and intrinsically unverifiable hypothesis which nevertheless would avoid the local intrinsic absurdities which the notion of self-intersecting world-lines implies. Despite his note of caution, there is no question that Einstein in 1949 was closer to the spatializing interpretation than two decades before.

This impression is only strengthened when we read Karl Popper's report on the conversation he had with Einstein shortly after the publication of Schilpp's volume:

I tried to persuade him to give up his determinism, which amounted to the view that the world was a four-dimensional Parmenidean block universe in which change was a human illusion, or very nearly so. (He agreed that this had been his view, and while discussing it I called him "Parmenides.") I argued that if men, or other organisms, could experience change and genuine succession in time, then this was real. It could not be explained away by a theory of the successive rising into our consciousness of time slices which in some sense coexist; for this kind of "rising into consciousness" would have precisely the same character as that succession of changes which the theory tries to explain away.[19]

Popper clearly saw that Einstein's leaning toward the Parmenidean 'block universe' and his commitment to classical determinism were two sides of the

same coin. But was it Einstein's last word? According to Carnap's testimony
a few years later his commitment to the static view seemed to be less firm
than it appeared to Popper:

Once Einstein said that the problem of Now worried him seriously. He explained that
the experience of the Now means something special for man, something essentially
different from the past and the future, but that this important difference does not and
cannot occur within physics. That this difference cannot be grasped by science seemed
to him a matter of painful but inevitable resignation.[20]

Carnap clearly was not happy about this statement since he suspected in
it the influence of Bergson — whom positivists in general not only dislike, but
do not even read attentively — and he was relieved when Einstein assured him
that it was not so. What then were the sources of Einstein's doubts? Was it
perhaps the result of Popper's proddings? Or of the paper which his disciple
Hans Reichenbach published at that time in which it was argued that the Now
is inescapable — even from the physical point of view?[21]

Two facts from the last year — in truth, from the last weeks — of his life
indicate that Einstein never ceased to struggle with the problem of time until
the very end. When he heard of the death of his friend Michel Besso, he
found consolation in the belief that death does not mean anything since for
physicists the distinction between the past, the present and the future is
nothing but a "stubborn human illusion." This was written on March 25,
1955, less than four weeks before his own death.[22] It would seem then that
Einstein died as a Spinozist and Laplacean, since his words are almost *verbatim*
the same as those used by Laplace who more than a century ago declared
that for a consistent determinist time is a mere incapacity to know everything
at once. Yet, was he entirely sure of it? At nearly the same time — certainly
not long before that date — Einstein wrote a foreword to the English translation of Louis de Broglie's book *Physique et microphysique* which had been
published in 1947, i.e., prior to the author's reconversion to classical determinism. It is a brief preface, hardly half a page; but what is interesting is that
half of it deals appreciatively with one essay out of the total number of
twenty-seven — 'Conceptions of Contemporary Physics and the Ideas of
Bergson on Time and Motion':

What impressed me most, however, is the sincere presentation of the struggle for a logical
concept of the basis of physics which finally led de Broglie to the firm conviction that
all elementary processes are of a statistical nature. I found the consideration of Bergson's
and Zeno's philosophy from the point of view of the newly acquired concepts highly
fascinating.[23]

To understand Einstein's interest we must at least briefly sum up the content of the whole of de Broglie's essay to which he referred. It was originally an article published in the *Revue de métaphysique et de morale* in 1941. But already in his earlier book *Matière et lumière*, in the essay, 'Les idées nouvelles introduites par la mécanique ondulatoire,' de Broglie pointed out that the impossibility of correlating a precise spatio-temporal location with a definite dynamic state of a moving body has a certain affinity with Zeno's famous paradoxes of motion; three years later he brought his reflection on Zeno and Heisenberg's principle in relation to Bergson's criticism of Zeno. Both Bergson and Zeno did not believe that motion consists of motionless points; but while Zeno concluded that motion cannot exist, Bergson concluded that motion is real; what is impossible is its reconstruction from static points. De Broglie then emphasized the fact that forty years before the formulation of the indeterminacy principle Bergson insisted rightly on the impossibility of motion *being* at any static point of its trajectory. In truth, borrowing the language of the quantum theories, Bergson could have said: "If one seeks to localize the moving object in a point in space, one will obtain nothing but a position, and the state of motion will completely escape." Since de Broglie wrote his essay at the time when he accepted the objectivist interpretation of Heisenberg's principle, he naturally welcomed Bergson's idea of 'weakened causality' according to which only the *general* character of the future, but not its specific details, are contained in the present – the view which is in agreement with the probabilistic interpretation of the laws of nature in contemporary physics.[24]

It is this that Einstein found fascinating. Mere polite words? This is extremely improbable since Einstein always could be very blunt when he strongly disagreed; or he could have simply ignored it entirely. Most probably the attentive reading of de Broglie's essay on Zeno and Bergson stirred in him some second thoughts about the adequacy of the traditional geometrization of time – the same doubts which he mentioned to Carnap – and perhaps even some fleeting doubts about his own life-long commitment to determinism. For there is no question that these two themes – spatialization of time and determinism – are, according to Meyerson, who is clearly following Bergson in this respect – closely related, being both two complementary manifestations of the same search for "identity in time."[25]

In the rest of his comment on *La déduction relativiste* Einstein praised Meyerson's opposition to phenomenalism and positivism and his insistence on *understanding* which can be achieved only when the experimental data appear as consequences within a deductive system: "The human mind is not content to posit relations; it wishes to *understand* . . . "[26] In this sense he classified Meyerson as a rationalist, although not in the sense of Kantian critical idealism. He pointed out that what separates Meyerson from Kant is his disbelief in the Kantian *a priori*; not even logical form and causality are necessary parts of our thought. (Whether this remark was compatible with Einstein's Spinozism is another question.) Finally, he was pleased with Meyerson's emphasis on the continuity of classical and relativistic physics, on the Cartesian overtones of general relativity, even though he added the warnings already quoted against the terms 'spatialization of time' and 'geometrization of matter.'

It is only fair to add that the differences between Meyerson and positivism were somehow exaggerated by both Meyerson and Einstein. Meyerson himself in *Identité et réalité* (p. 511) regarded what he called the positivistic 'principle of legality' as a mere abbreviation of the principle of causality. Jean Piaget pointed out that the difference between Meyerson's concept of causality and, for instance, that of Philipp Franck is that of emphasis rather than of substance; in the latter case the search for identity in time is merely semantically disguised.[27] The same attitude is discernible in most members of the Vienna circle; their commitment to classical determinism was, despite some terminological modifications, quite strong. When Niels Bohr and Pascual Jordan dared to draw wider philosophical consequences from the revision of the principle of causality in modern physics, they were attacked, especially the latter, with some passion by all of them with the single exception of Reichenbach.[28] Even the search for identity in space was not absent: Moritz Schlick spoke of 'the truly magnificent enterprise' (*ein wahrhaft grossartiges Unternehmen*) to dissolve all various phenomena of nature into the all-pervading aetherial fluid and compared it to Thales' formula that 'All is water.'[29]

Some physicists may be disappointed that Meyerson's book does not contain a single mathematical formula; but in this respect, it is no exception in the literature dealing with philosophy of science. Thus Bridgman's *The Logic of Modern Physics* does not contain any formula either; Reichenbach's *The Philosophy of Space and Time* only a few. Still, Meyerson could have strengthened his arguments by at least a brief discussion of the Lorentz transformation and Minkowski's formula for the constancy of the world interval. It is true that he refers to Langevin's analysis of the latter formula, but only in his second book (*E.S.*, II, 183, 377), not in this book; Langevin's analysis should have helped him to point out the error of G. Moch who claimed that although the order of causally related events cannot be reversed (which is true), it can for certain observers degenerate into simultaneity (which is false).

But in this oversight (§50 of the text below) Meyerson is not alone; a similar error was made not only by Jean Piaget, but also by Philipp Franck.[30] He also diplomatically avoids a discussion of the famous 'twin paradox'; Bergson's controversial book *Durée et simultanéité* is mentioned only in the preface and designated as 'the masterful work' (*l'oeuvre magistral*) which is certainly true only *cum grano salis*. Meyerson did not mention the discussion which took place between Bergson and André Metz in 1924, i.e., between the publication of *Durée et simultanéité* and *La déduction relativiste*; this was probably just as well since errors were committed by both disputants and because it was not easy to disentangle what was correct from what was wrong in Bergson's critical comments on relativity; this is why the controversy on this point was not dead even after half a century.[31]

After *La déduction relativiste* Meyerson wrote only two books: *Du cheminement de la pensée* (1931) and *Le réel et le déterminisme dans la physique quantique* (1933). Of the second, he saw only the proofs; he died before its publication. The first voluminous book – three volumes and nearly one thousand pages – consisted of, besides the restatement of his main theses, also a chapter on primitive mentality which, according to him, does not basically differ from the modern mentality even though the identifying thought patterns are of a much cruder kind; also of his analysis of traditional and modern logic. It was not difficult for him to show the importance of identification in the syllogistic of Aristotle as well as in modern symbolic logic ('la logistique' as it is called on the continent) in which tautology has such an important role. In the history of logic, Antisthenes, who insisted on the absolute identity of the subject and the predicate and on the impossibility of other judgments than identical, is a counterpart of Parmenides in the history of ontology. In the same way as the Eleatic ideal of a single and immutable Being remained unattainable, a complete reduction of logical relations to tautology has never been completely achieved; another instance of the resistance of the surd 'irrational' element to the identifying tendencies of the intellect.

This also explains Meyerson's rather open-minded attitude toward recent microphysical theories. There is no question that classical physics yielded far more easily to Meyerson's interpretation of science than the physics of the twentieth century. Meyerson claims – and he is certainly not alone – that the physics of relativity is a continuation and even a culmination of classical physics, that the concept of permanent substratum was only modified, but not given up, since two separate conservation laws of classical physics were merged into one single law of conservation of mass-energy (see §181 below).

Thus, according to him, the difference between relativism and mechanism is merely that of degree. On the other hand, the facts on which the theory of quanta and, in particular, the wave mechanics of de Broglie and Schrödinger were based do not fit the substantialistic interpretation to which classical physics yielded so beautifully; the substitution of statistical laws for causal laws, a surrender of classical causality of the Laplacean type, were real and radical departures from the mechanism of the past centuries. But Meyerson is ready to concede it; he regards the indivisibility of Planck's constant of action h, which is a root cause of so many upheavals in the physics of this century, as a new kind of irrationality which his epistemology in principle – though not in its concrete form – fully anticipated. It is understandable that in Chapter VI, "L'irrationnel," of his first book, quanta are not mentioned at all; this takes place only in his second book in Chapter VI, with the same title as in the first book, while more frequent references are in *La déduction relativiste* §§113–114, then in §127 when he speaks of Bohr's and Sommerfeld's models of the atom; finally in §286 he recalls the view of Émile Borel on the possibility of abandoning the concept of spatial continuity in the domain of quantum phenomena.

Yet, one can see that despite his open-mindedness, he feels rather ill at ease in the presence of quantum physics. This is visible in his last two books – in *Du cheminement de la pensée* and, in particular, in *Le réel et le déterminisme dans la physique quantique*. He retains a noncommittal and undogmatic attitude toward microphysical indeterminism, even though he finds only very few of its historical antecedents – Epicurus, Lucretius, Renouvier and Boutroux; he is apparently unaware of C. S. Peirce and, in this matter, Bergson. But he definitely thinks that the present departure from intuitive models in microphysics is merely temporary; he quotes approvingly (p. 19) the requirement of *imaginability* of physical models which John Tyndall set up as the main epistemological criterion for the physicists in his Liverpool address in 1870; he quotes approvingly an early statement of Arthur Compton that "if there are waves, there must be the medium in which they are propagated." This shows how deeply was his mind immersed in the classical modes of thought, although, in fairness to him, one must admit that he was not – fortunately – entirely consistent in this regard: what is less intuitive (in a Tyndallian or Kelvinian sense) than non-Euclidean geometry, which he accepted? Still, on the whole, one must agree with Gaston Bachelard when he wrote that

in the presence of the surprising principles of quantum mechanics, Meyerson himself, who spent the treasures of meditation and erudition in order to prove the classical character of relativity, suddenly hesitates. One can doubt that he will ever write *Déduction quantique* to complete the demonstration undertaken in *La déduction relativiste*.[32]

This will finally lead us to one question which Meyerson's epistemology failed to answer. It is difficult to deny that he succeeded in establishing convincingly and beyond any doubts by accumulating overwhelming evidence that the process of explanation in all periods of Western thought in both philosophy and sciences consisted in attempts to eliminate diversity in both space and time. At the same time, he showed – equally convincingly – that such a search for identity has never fully succeeded and also cannot ever fully succeed since its very success would result in the most extreme form of static monism, i.e., in a complete negation of experience. This is a rather frustrating situation and Meyerson is aware of it when he speaks in Chapter XVIII of his *De l'explication dans les sciences* of the 'epistemological paradox.' Yet, the interplay of these two opposite tendencies remains a fact and they are present in different degrees in every scientist, even in those who because of their lack of interest in – as well as the lack of knowledge of – the history of science and history of philosophy tend to dismiss Meyerson's theory of science as a mere metaphysics. But his philosophy could be far more convincing and would thus cease to be paradoxical, if it could be shown *why* and *when* the identifying tendencies succeed in order to differentiate such circumstances from those when they fail.

To this question Meyerson does not provide any answer, although, in my view, it *can* be given. We have only to remember that the most – in truth nearly all – of Meyerson's illustrations of the identifying process are taken from either the history of philosophy or from *classical* sciences, especially from physics, in which the idea of substance in various forms had such a decisive role. Whether such entities were material or mental, whether they were individual or all pervading, whether they were the atoms of Democritus or of Dalton, the monads of Leibniz or the 'reals' of Herbart, whether it was the plenum of Descartes, or the substance of Spinoza, or the Absolute of Neo-Hegelian idealists – they shared one feature: they were *substantial* entities either persisting in time or even beyond time altogether. This, let me repeat it, Meyerson established beyond any doubt: his documentation, no matter how lavish, could still be increased by more recent examples: let us remember, for instance, the rarely noticed affinity of Russell and Herbart, the similarity of the contemporary 'mind-dependent theories of becoming' with the Eleatic philosophy, etc. We know that in classical physics this pattern of thought led to the triumphs of the corpuscular-kinetic models and to the establishment of Laplacean determinism; also that the same patterns of thought failed when they were applied beyond the limits that Reichenbach called "the zone of the middle dimensions."

This, indeed, is the most striking feature of the physics of the twentieth century: a radical departure from the intuitive mechanistic models of the past. The association of the undulatory and corpuscular natures of the microphysical entities can be *visualized* (in the Tyndall-Kelvinian sense) as little as the finite and limitless space of Riemann. Meyerson was once accused by one positivistic critic of a substantialistic bias;[33] but it was unfair to call it 'bias' since he viewed the concept of substance as a historical outcome of the search for identity in time. But it is easy to imagine his embarrassment and disappointment at the sight of the newly discovered microphysical entities which are still called 'particles' only because of the inertia of our language and of our imagination in spite of their fleeting and almost vanishing duration. It is true that he would probably again speak of another irrationality; but does not then his *'l'irrationnel'* become a mere convenient escape-route which is used whenever the facts do not fit his analysis? Is not such an escape route merely a disguised impasse?

The way out of this impasse is, I am convinced, opened by the *evolutionary* or *genetic theory of knowledge*, advocated by such otherwise different thinkers as Reichenbach, Bergson and Piaget.[34] In the light of this theory and, in particular in the light of Piaget's researches, unknown to Meyerson, the concept of the object persisting in time is formed in very early childhood under the pressure of macroscopic experience. It is then hardly surprising that such a concept does not apply outside the realm of middle dimensions where it is not only useful, but even vitally important; Elie Zahar rightly insists against Meyerson on its survival value.[35] Is it not then natural that such a concept, being a result of our cognitive adaptation to limited strata of reality, fails beyond its limits? Thus the electron *cannot* be a micro-object differing only in its dimensions from the solid bodies of our sensory experience; nor is the universe a sort of 'mega-object,' as Piaget pointed out. (He could have added an additional decisive argument based on modern cosmology: the expanding time-space is not an 'object' in our sense since the macroscopic objects of our sensory experience consist of *simultaneous* parts; but the concept of simultaneity loses its meaning on the scale of huge distances and extremely large velocities such as those of fleeting galaxies.) In other words, what Bergson called the 'logic of solid bodies' which is a result of the adaptation of our intellect to the zone of the middle dimensions, ceases to be applicable to the microcosmos as well as to the megacosmos.

The evolutionary theory of knowledge implies *the evolution of reason* in the sense that new forms of thought irreducible to the traditional categories do emerge under the pressure of new experience. This appears to be altogether contrary to Meyerson's insistence on the *immutability* of human reason as we find it especially in the concluding chapter of *De l'explication dans les sciences*. It is true that he cautiously concedes that the opposite view is not impossible, but he eventually rejects it as a 'brilliant' and even 'dangerous' heresy. He regards Hegel's failure to deduce becoming (which he analyzed extensively in the same book) as a confirmation of the immutability of reason; despite Hegel's effort, becoming remained irrational after him as it was before, and even disappeared in the philosophy of one of his most faithful disciples. Obviously, if reason is essentially immutable, what appears irrational now will remain irrational forever – a rather discouraging and pessimistic conclusion which was furthermore disproved by the evolution of the sciences. It is hardly a century ago that Herbert Spencer still regarded Euclidean geometry as the only conceivable one and there are even today philosophers who regard any departure from classical determinism as a 'suicide of reason.' In his critical survey of modern logic in *Du cheminement de la pensée* Meyerson characteristically does not mention three-valued logic; he who was so open-minded toward a thorough revision of the foundations of physics and geometry excluded apparently more radical departures from traditional logic. For him Eleaticism was a 'hereditary sin' – or would he prefer to call it a 'hereditary virtue'? – of human reason. As a mere statement of fact, as a mere diagnosis, he was unquestionably right. It is also only fair to say that in his later books he, at least occasionally, conceded that it was just that – a diagnosis, not a prognosis. But one can find only few books in which such diagnosis is equally penetrating and equally thoroughly documented.

MILIČ ČAPEK

NOTES

[1] José Ortega y Gasset, *The Revolt of the Masses* (Mentor Books, 1950), p. 80.
[2] H. T. Costello in his review of the first three of Meyerson's books in the *Journal of Philosophy* 22:637.
[3] H. Samuel, *Essay in Physics* (Oxford: Blackwell, 1951), p. 70.
[4] F. Enriques, *La dottrine di Democrito d'Abdera* (Bologna, 1948), Ch. III; Carl Liepmann, *Die Mechanik der Leucipp-Demokritischen Atome* (Leipzig, 1885).
[5] *De rerum natura*, II, v. 294.

[6] *Immanuel Kant's Critique of Pure Reason*, tr. by Norman Kemp Smith (MacMillan: London, 1953), p. 215.

[7] *De l'Intelligence* (Paris: Hachette, 16th ed.), p. 11.

[8] Kurd Lasswitz, *Geschichte der Atomistik* (Hamburg-Leipzig, 1890), 2:102.

[9] ES 1:162.

[10] S. Sambursky, *The Physical World of the Greeks* (MacMillan: New York, 1956), pp. 202–3.

[11] *Bulletin de la société française de philosophie*: séance du 6 avril 1922, p. 107ff., trans. on pp. 257ff. below.

[12] Bergson was one of the first who recognized the significance of Meyerson's work: cf. his 'Rapport sur *Identité et réalité* d'E. Meyerson.' *Séance et Travaux de l'Académie des Sciences Morales et Politiques*, 23 janvier 1909.

[13] Meyerson quotes the secondary source: Marcolongo, *Relatività* (Messina, 1921), p. 98. The original of Einstein's sentence is in Appendix II of his book *Relativity. The Special and General Theory*, tr. by Robert W. Lawson (Crown Publishers: New York, 1961), p. 122: "From a 'happening' in three-dimensional space, physics becomes, as it were, an 'existence' in the four-dimensional world."

[14] 'A propos de *La déduction relativiste* de M. Émile Meyerson,' trans. André Metz, *Revue philosophique de la France et de l'étranger* 105 (1928) 164–165, translated from the French by David A. and Mary-Alice Sipfle (italics mine). Cf. p. 255 below.

[15] G. Boas in his review of the English translation of *Identité et réalité* in *Journal of Philosophy* 28 (1931) 19.

[16] A. Eddington, *The Nature of the Physical World* (New York: MacMillan, 1933), p. 137.

[17] *Albert Einstein: Philosopher-Scientist*, ed. by P. A. Schilpp (Evanston, Ill., 1949), pp. 557, 560, 687–88.

[18] Pierre Duhem, *Le système du monde* (Paris, n.d.), 1:168–9.

[19] *The Philosophy of Karl Popper*, ed. by P. A. Schilpp (Open Court:La Salle, Ill., 1974), pp. 102–104.

[20] *The Philosophy of Rudolf Carnap*, ed. by P. A. Schilpp (Open Court: La Salle, Ill., 1963), pp. 37–8.

[21] H. Reichenbach, 'Les fondements logiques de la mécanique des quanta,' *Annales de l'Institut Henri Poincaré* 13 (1952): 156 [English transl. in *Hans Reichenbach Selected Writings* 2:237–278, ed. Maria Reichenbach and R. S. Cohen (Reidel: Boston and Dordrecht, 1978)].

[22] *Albert Einstein–Michel Besso Correspondence 1903–1955*, tr. by Pierre Speziali (Paris: Hermann, 1972), pp. 538–39.

[23] L. de Broglie, *Physics and Microphysics*, tr. by Martin Davidson (Harper & Brothers: New York, 1955), p. 7.

[24] It is important to stress that the original essay 'Les conceptions de la physique contemporaine et les idées de Bergson sur le temps et sur le mouvement' was on the wish of Louis de Broglie himself – who meanwhile returned to determinism – not translated, but merely summarized! This is why it is important to read it in the original.

[25] Most recently, Karl Popper claimed on the basis of two letters of Wolfgang Pauli to Max Born in 1954 that Einstein eventually gave up his belief in the deterministic four-dimensional 'block-universe.' *Abstracts of the Seventh International Congress of Logic, Methodology and Philosophy of Science at Salzburg, Austria* (1983, Vol. IV, p. 176).

But this view is difficult to reconcile with Einstein's letter on March 25, 1955 quoted above. The only thing that the two Pauli letters prove is that Einstein was more concerned about epistemological realism than about determinism, not that he rejected determinism definitely and unambiguously. But very probably he had occasionally very serious doubts about it as his preface to the translation of de Broglie's book indicates. The manuscript of this preface was, indeed, written in 1955, i.e., in the early months of the same year. (I am indebted for this information to Professor John Stachel of the *Einstein Project* at Princeton University Press and Boston University.)

[26] 'A propos de *La déduction relativiste* de M. Émile Meyerson,' trans. André Metz, *Revue philosophique de la France et de l'étranger* 105 (1928), 164, translated from the French by David A. and Mary-Alice Sipfle. Cf. p. 254 below.

[27] Jean Piaget, *L'Introduction à l'épistémologie génétique* (Paris: Presse Universitaire de France, 1974), 2:296—300.

[28] On the views of Bohr and Jordan: E. Meyerson, *Le réel et le déterminisme dans la physique quantique* (Paris: Hermann, 1933), pp. 27—28; M. Čapek, *Bergson and Modern Physics, Boston Studies in the Philosophy of Science*, Vol. VII (Dordrecht: Reidel, 1971), Appendix II ('Microphysical Indeterminacy and Freedom').

[29] *Grundzüge der Naturphilosophie* (1948), p. 88 [Eng. tr. *Philosophy of Nature*, N.Y., 1949].

[30] J. Piaget, *The Child's Conception of Time*, tr. by A. J. Pomeranz (Ballantine Books: New York, 1969), p. 305. On Philipp Franck's similar error cf. M. Čapek 'The Inclusion of Becoming in the Physical World,' in: *The Concepts of Space and Time, Boston Studies in the Philosophy of Science*, Vol. XXII (Dordrecht: Reidel, 1976), p. 523.

[31] On this point cf. Marie-Antoinette Tonnélat, *Histoire du principe de relativité* (Paris: Flammarion, 1970), pp. 280—293; M. Čapek, 'Ce qui est vivant et ce qui est mort dans la critique bergsonienne de la relativité,' in *Revue de Synthèse*, No. 99 (1980), pp. 313—344.

[32] Gaston Bachelard, *Le nouvel esprit scientifique* (Paris, 1934), 4th ed. (Presse Universitaire de France, 1946), p. 175.

[33] Albert E. Blumberg, 'Emile Meyerson's Critique of Positivism,' *The Monist* 42 (1932) 60—79.

[34] M. Čapek, *Bergson and Modern Physics*, Part I, pp. 3—80.

[35] Elie Zahar, 'Einstein, Meyerson and the Role of Mathematics in Physical Discovery,' *British Journal for the Philosophy of Science* 31(1980) 1—43, esp. p. 11.

THE RELATIVISTIC DEDUCTION

PREFACE

This book is closely related to our earlier works, *Identité et réalité* (2nd ed., Paris: Alcan, 1912) [*Identity and Reality*, trans. Kate Loewenberg (New York: Macmillan, 1930] and *De l'explication dans les sciences* (Paris: Payot, 1921). In the same way as we examined analogous concepts in those works, here we shall study the concepts of Einstein and his continuators from a strictly limited point of view, attempting to expose the thought processes used by these scientists and their supporters.

We are not unaware that such an undertaking presents special difficulties. Some critics who have otherwise favorably received our works have seemed to suggest that we devoted too much time to presenting and analyzing outdated scientific theories. However, the study of such theories offers an important advantage: since these hypotheses have now been discarded, we are sure to approach them without prejudice, which makes it easier to discover the underlying mechanism of thought. In the case of theories that are still fighting for their existence, the fact that we are personally involved — because scientific beliefs are part and parcel of man's deepest intellectual being — almost forces us to be partial in one way or another. Thus the reader should bear in mind that it would be futile to expect a contemporary of Einstein, Langevin, Weyl and Eddington to be able to escape their influence entirely and view the theory of relativity with as much detachment as he would the hypothesis of phlogiston.

If we have, nevertheless, decided to make the attempt, it is because we consider the task particularly interesting from the standpoint of the scope of the ideas we set forth earlier. In fact, some of the critics who thought we overused the history of science also seemed to believe that our analysis of outmoded doctrines led us to a concept of science that might well have characterized the past, even the very recent past, but in no way applies to contemporary physics. The merest hint of such an argument is apt to have a particularly strong influence on the reader's mind, which is already prepared to accept it. Indeed, the exuberant growth of modern science has almost inevitably led to the opinion that this process involved a way of thinking essentially different from that of preceding generations, and there is no doubt that the powerful influence of Auguste Comte has contributed to establishing

3

the belief that the *positive* science of today stands in clear contrast to the *metaphysical* lucubrations of our forefathers. Moreover, this turn of mind is not peculiar to our epoch; at the height of the Middle Ages, in the thirteenth century, Roger Bacon proudly said, "we moderns" (*nos moderni*), and before Lavoisier the term 'modern chemistry' was in general use among phlogiston theorists. Convictions of this sort have some value insofar as they motivate the scientist to distrust tradition and to open new avenues of thought. But we believe the epistemologist would do better to take his inspiration from Pascal's thought that

the whole succession of men throughout the course of so many centuries is to be taken as one single man who lives on and on and learns continuously.

This idea of the solidarity of men of all times in the effort to discover truth is more apt to reveal to us the motives that actually govern the intellect. We believe the present study will confirm this, for we hope to convince the reader that relativity theory, so modern and at first glance so strange, fits perfectly the pattern we saw emerge from our earlier examination of other concepts, some of which might appear to belong to quite a different order of thought.

Given the particular nature of the task we have set ourself, we felt justified in setting aside almost completely an entire aspect of the relativistic concepts with which philosophers have been much concerned, namely, their more properly psychological or subjective aspect, dealing, for example, with the relationship between the Einsteinian and Bergsonian notions of time — a question, moreover, that Bergson himself has treated quite recently in a masterful work.[1] It is not that we meant for a single moment to underestimate the great value of this line of research; it is simply that it seemed possible to rule out such a study without damaging our presentation. Although we shall seek to elucidate questions concerning the psychology of scientists, we prefer to do so by studying the objective aspect of the theories they conceive.

We must likewise warn the reader that he will not find here any attempt to present relativity in the language of common sense and thus make it accessible to the general audience of educated men unacquainted with mathematics. We dare say, on the contrary, that such an enterprise is doomed in advance — a point we shall develop later (§56 ff.). The most we can do to make the ultramodern theories seem less forbidding is to compare them with certain concepts from the past that are more familiar to the philosopher.

We aspire even less to argue with the relativists about the foundations of their theories, or even about any of the essential features of the theories. Relativity is a scientific system arising out of certain experimental observations: competent physicists (or at least a great many of them) believe that these observations can be accounted for, made to fit into the *totality* (to use Höffding's apt expression), only by modifying this concept of totality (and more particularly the concepts of time and space) in a particular way. Consequently, it is clear that the fate of the theory will depend above all on these very observations.

Certainly, the experimental bases of relativity theory are not nearly as narrow as its adversaries sometimes claim; the reader will see (§76) that there is a close connection between relativistic concepts and the body of beliefs about electricity adopted by physicists following Maxwell, Hertz, and H. A. Lorentz. It is obvious, however, that the electrical theories alone had not been sufficient to give rise to relativism, which, as we know, owed its origin to the Michelson—Morley experiments and its subsequent success to the fact that the results predicted by Einstein were confirmed by observation. If, therefore, the impossible should occur and the results of the 1919 and 1922 eclipses should be contradicted by later observations, Einstein would probably lose most of his supporters. If it were later shown that the spectroscopic phenomena he predicted (just as he predicted the phenomena that were to be observed during an eclipse) did not take place, he would have very few believers left. The last of these would no doubt abandon him if new astronomical facts should be discovered capable of accounting for the anomaly of Mercury, or electrical phenomena able to explain the paradoxical observations of Michelson and Morley in some way other than by relativity. On the other hand, if all these observations were able to be applied in new areas to explain phenomena that have continued to be puzzling (as has just happened in the case of the Bohr atom, where Sommerfeld has succeeded in accounting for fine spectral lines by introducing the concept of relativity into the motions of the electron (cf. §127)), the last opponents would certainly be reduced to silence.

This, of course, does not imply that one cannot criticize the way in which the physicist has connected his observations and his theory, that is to say, how he has deduced the latter from the former. But one must realize that such criticisms, even if perfectly justified, are very unlikely to have any effect on him. For he *must* have a theory, a concept of totality tying all his observations together, and if relativity seems sufficient to the task, one can be sure (we have pointed this out many times) that he will abandon it only if someone offers him a better theory, perhaps less paradoxical in appearance, but above all (for this is his principal concern) more comprehensive, in better agreement with the observations, explaining a greater number of them.

Moreover, we must keep in mind, as we also stressed in *De l'explication* (ES 2:360),[2] that scientific knowledge has the advantage over philosophical knowledge in terms of its solidity. The philosopher himself cannot avoid making this observation if he but considers the evolution of these two branches of human knowledge. Kant had already said:

One can easily guess which side will have the advantage in the battle between two sciences, one of which surpasses all others in certainty and clarity, while the other is only striving to acquire these qualities.

It is true that in this passage Kant was comparing philosophy with mathematics, but it is well-known to what extent modern physics is permeated with mathematics, and there is no need to underline the fact that relativity theory represents a sort of *summum* in this respect.

Nevertheless, since science and philosophy are only two different ways in which human reason attempts to grasp reality, they necessarily meet at many points. This juxtaposition ought to instill in the philosopher an appropriate distrust of his means of verification; he must be thoroughly convinced that the scientist comes under his jurisdiction only when he ventures into deductions or lines of reasoning going far beyond the observations. As the reader will see, this is what sometimes seems to have happened in the case of at least some protagonists of the new theory.

But, in general, we shall consider relativity as a particular phase of physics, a phenomenon that took place at a particular moment in the evolution of science. Thus, the validity of our arguments will not depend on the vicissitudes of the doctrine of relativity itself. Even if the entire edifice of the Einsteinian theory should crumble tomorrow, even if the whole superb construction should disappear from physics without a trace, it will nonetheless remain true that it existed, that its authors reasoned in a given way, that their arguments appeared convincing to the majority of scientists competent to judge them (although this adds no great weight to our argument), and that it is therefore legitimate to draw inferences about the principles of scientific reasoning in general from an analysis of these deductions.

Because of the close relationship of this work to our previous publications, we necessarily cite them frequently. This is a disadvantage we have not been able to avoid, and we beg the reader's indulgence. We have, however, done our best to set forth here the essentials of the concepts in question, so that the reader can, if need be, forgo all reference to our earlier writings. But for this reason we have had to reproduce a certain number of citations that we have already used elsewhere, lest we give the reader the impression of having made unsubstantiated claims. By this avowal we hope to mitigate the offense.

This book contains many a page to which we cannot claim exclusive rights. First of all, the original idea of such a work arose out of a conversation with Paul Langevin on the eve of Einstein's arrival in Paris. We were led to observe how much Langevin's position on the realistic nature of relativity theory — a position at which he had arrived independently — agreed with the principles we believed could be deduced from an examination of the physical sciences in general and their historical development in particular. Langevin also provided us with some of our documentation (which unfortunate circumstances beyond our control have prevented us from making as complete as it should have been, and for which we also apologize) and constantly helped us overcome the technical difficulties that arose. We cannot adequately express our gratitude. We also owe much to André Lalande and Dominique Parodi, who read the work before it was completed and made observations serving as the point of departure for material incorporated into the book in its present form.

Leysin, March 1924   EMILE MEYERSON

NOTES

[1] [*Durée et simultanéité* (Paris, 1922), trans. Leon Jacobson, *Duration and Simultaneity* (Indianapolis: Bobbs-Merrill, 1965).]
[2] [See Abbreviations, p. xix above.]

CHAPTER 1

THE QUANTITATIVE

1. *The Role of Mathematics*

As the title of this work indicates, we shall be concerned with the theory of relativity insofar as it constitutes a body of deductions. In relativity, as everyone knows, these deductions are made by means of mathematics. It therefore seems useful, before considering relativity theory itself, to say something about the role of mathematics in physical theory in general.

This role, needless to say, has grown steadily in modern physics. Attempts to the contrary have not had the slightest success. The latest of these efforts at extramathematical deduction — those of Goethe, Schelling and his school, and Hegel — have left no noticeable trace in science, which has not even bothered to refute them, but has simply brushed them aside and ignored them.

2. *How Positivism Explains this Role*

Attempts have been made to explain the predominance of mathematics in modern science simply in terms of its usefulness in providing the most precise kind of description. "I often say," declared Lord Kelvin, "that when you can measure what you are speaking about and express it in numbers you know something about it; but when you cannot measure it, when you cannot express it in numbers, your knowledge is of a meagre and unsatisfactory kind."[1]

3. *The Inadequacy of this Explanation*

This way of justifying the concern for *measurement* — which science does indeed impose inexorably on all phenomena daring to enter its domain — solely in terms of lawfulness and predictability can be seen to be consistent with the positivistic prejudice that these are the only considerations that would have sufficed to give science its present form. But here, as on many other points, the justification is clearly insufficient. While it is true that mathematical description — measurement — does permit us to describe one aspect of a phenomenon more precisely (cf. IR 380 ff.; Eng. 341 ff.),[2] it is certainly unsuitable for capturing other aspects of it, notably anything that has to do with quality. This constitutes a more or less voluntary sort of sacrifice, which cannot be overemphasized, because it may serve to enlighten us as to the true goal of science.

4. *The Importance of Quality*

Nothing is more certain than the fact that science takes its point of departure from common sense perception and that the reality presupposed by this perception is a mixture of the quantitative and the qualitative. Undoubtedly the 'immediate data' of our sensation, as Bergson made clear by his profound analyses, can only be entirely qualitative. These sensations are fleeting, however, as is amply demonstrated by the very fact that an extremely subtle search was needed to discover them. What our consciousness furnishes in their place is actually *perceptions*, the result of a complex and unconscious process of reasoning, for which the immediate data served as the point of departure. Clearly this process is already an evolution toward the quantitative. Indeed, common sense objects do appear to us in space and as occupying space, thereby coming under the category of quantity. However, at the same time we affirm that they are endowed with qualities, and it is certainly not at all correct, as physics may have accustomed us to think, that the quantitative element initially appears most important to us. On the contrary, from the point of view of action alone, it is clear that the qualitative is given precedence over the quantitative. In fact, in the great majority of cases quantity can interest us only *after* we have ascertained the quality. There is no need to elaborate on our use of this principle in the case of our most urgent need, the need for food. This is equally obvious at what is, in some sense, the other end of the hierarchy of values, when we say we would give the complete works of some fashionable novelist for one page of Anatole France.

5. *Quality and Action*

It would not even be accurate to say that it is quantity rather than quality that constitutes the proper ground of our action. Certainly animals in general (with the possible exception of species such as the ants or the bees that impress us as having a quasi-human attitude in this respect) simply use what they find in nature just as they find it. Man, however, as soon as he advanced beyond the brute animal stage, set out to transform what nature spontaneously provided, to modify the form, that is, the properties, of the stone, the horn or the piece of wood in order to make them usable, to soften meat mechanically and eventually by fire in order to make it easier to digest. It is precisely this will to transform the properties of things, including the use of tools born of this tendency, that has generally, and certainly justifiably, been seen as the most characteristic criterion of incipient humanity, the trait distinguishing it from animality. Now in examining scientific and philosophic concepts we obviously find ourselves in a domain belonging exclusively to man. It is true that, in a sense, the world of perception (probably almost identical in man and animal) is only an incipient form of scientific and philosophical system. Thus, from whatever angle we approach the problem, we are led to the conclusion that, from the point of view of our action on the real world, it is actually the qualitative that should first attract our attention.

6. *Quantity and the Nature of Things*

Nevertheless, as soon as we begin to reflect on the true nature of reality, it is certain that the quantitative is apt to appear more important than quality. Although there have, of course, been attempts by science to formulate qualitative theories (Aristotelianism in antiquity and the Middle Ages, for example, or, quite recently, the extramathematical hypotheses mentioned above), it is nonetheless true that from the beginning of scientific thought, with Leucippus and Democritus, there has been a powerful conviction that quality does not belong to the innermost nature of a body, but is only an 'opinion' added to this nature, which consequently consists exclusively of quantity. This way of seeing things was expressed quite rigorously by Descartes:

The nature of matter or of body in its universal aspect, does not consist in its being hard, or heavy, or coloured, or one that affects our senses in some other way, but solely in the fact that it is a substance extended in length, breadth and depth.[3]

To be sure, Descartes's ideas met with some resistance from his contemporaries (due mainly to the fact that for centuries all human knowledge had been dominated by an opposing theory, that of Aristotle, which put quality foremost), but his position very quickly prevailed. Soon afterward, Locke, starting from a rather different position, introduced his distinction between primary and secondary qualities, the former referring exclusively to space and the occupation of space, thereby reinforcing the conviction that the quantitative is of primary importance in the real world. This has become a sort of inviolable axiom for the modern physicist. Kelvin's remarks have been quoted above. Planck stated the *credo* unambiguously: "That which can be measured is by that very fact real." Von Laue went still further, affirming that "what in principle disallows all measurement is certainly not real from the physical point of view." [4] Bergson has said that

the physicist has the right and the duty to substitute for the datum of consciousness something measurable and numerable with which he will henceforward work while granting it the name of the original perception merely for greater convenience. [5]

One could not better characterize the fundamental tendencies of modern physics and, at the same time, its relation to sensory reality. These well-known characteristics of the evolution of modern thought, in conjunction with the appearance of analogous ideas in the Greek atomists, lead us to suspect that we must be dealing with an attitude that our intellect has suggested to us almost spontaneously from the very beginning of its relationship with nature.

7. Change and its Explanation

Our sensations are constantly changing, and we must understand, that is, deduce, this change. Of course, we could also deny it entirely, declaring that reality is immutable and undifferentiated because it cannot be anything else, given that only what is identical to itself in time and space conforms to the exigencies of reason, that is to say, is really intelligible: this is the sphere of Parmenides, a concept that abolishes all the reality of sensation in a single stroke. But here we have an affirmation that, by its brutal frankness, certainly shocks our deepest sensibilities; furthermore, by making phenomena disappear, it obviously has the fundamental disadvantage of rendering all science impossible, or rather superfluous, absurd. Consequently we would be very glad to reinstate, to 'save' (to use the well-known Greek term) changing reality. We must therefore seek a way to reconcile identity and diversity, which can only be accomplished by connecting the second of these terms to the first by a purely mental process, a deduction. Reason can accomplish this in two different ways. First of all, it can preserve, insofar as possible, the object as common sense presents it to us, which obviously implies that we must conserve its qualitative aspect. But in that case we have to explain how these qualities come to change. One can, for example, suppose that the qualitative element remains what it is but simply moves elsewhere (this is the 'substantial form' of scholasticism). One can also seek to establish that qualities, analogous but opposite properties, appear in an undifferentiated medium, making it possible for each one separately to be markedly different even though their sum remains the same. Finally, one can imagine still other devices tending to explain how one quality, while remaining fundamentally what it is, can nevertheless appear to be changed into another. One need only read through the series of concepts of quality mentioned above, from Aristotle to Hegel, to realize that this is what they are trying to do, and from this point of view the followers of Aristotle were not completely wrong in insisting that his views conform to common sense.

8. *The Artistic Point of View*

One might add that this is also the point of view of art, which, in its highest forms, is actually seeking to embrace the whole of sensible reality and to render it as completely as possible without losing any of it. For example, impressionism's contribution to painting was to bring out and fix nuances of sensation, nuances that are undoubtedly very real, since even the public at large has finally recognized this reality, but are at the same time so ephemeral that they disappear in the image habitually presented to us by our perception. Goethe's position shows us the same thing. The accusations he launched against Newton were certainly not only unjust; they were literally senseless, to the point that even German scientists, in spite of the unlimited — and otherwise legitimate — respect the great figure of their national poet always inspired in them, refused to take them seriously. But one can see how such a man, almost intoxicated by all the splendor of nature, must have been troubled by the mathematical approach, which, while pretending to seize the whole of reality in all its diversity, is willing to overlook the very part the artist considers the most essential. In a recent work on the relationship between science and philosophy, the famous physicist Wien, while firmly maintaining the complete accuracy of the scientific point of view, nevertheless quite rightly pleaded extenuating circumstances in Goethe's case (WW 221).[6]

9. *Conflicts and their Resolution*

We have here a universal and profound conflict, one of the many contradictions in which our intellectual being is steeped, for each one of us, however imbued with the principles of science, conceals within himself something of Goethe's state of mind. All of us, unless we are altogether closed to the testimony of either science or art, feel that they embody divergent tendencies. Similarly, we have the feeling that when we try to explain reality exclusively in terms of matter — as science necessarily does — we are attempting to do something in which complete success is impossible, since thought too is part of reality and is clearly left out of such an interpretation. To be sure, the immediate, naive common sense conception succeeds in bringing these disparate components together, but each of them is included only in a very rudimentary form; moreover, the resulting compound itself is not very logical. However, as soon as we begin to reflect on the nature of each particular branch of our mental activity, especially when we try to make the union between them a little less superficial, conflicts arise and these activities establish themselves in independent, and in many respects opposing, domains. In the course of our daily lives we are forced to effect a sort of reconciliation, maintaining the laws of common sense as far as we can, sometimes even in spite of what we know to the contrary; the rest of the time we allow either one or the other of the opposing points of view to prevail. We keep hoping, nevertheless, to overcome this syncretism and arrive at a true synthesis; only such a synthesis seems to us to be capable of representing reality in all its fullness. That is why the philosopher who promises to perform this miracle is — and no doubt always will be — certain to seduce us. But however attractive such a concept may be, and even if it should succeed in reconciling these contraries immeasurably more completely than any system thus far offered to us, it must be realized that it would have no effect whatsoever on the course of science. For science, by a postulate that may be implicit but is nevertheless absolutely fundamental, has made its choice once and for all, giving up quality in favor of quantity. And it is not too difficult to recognize what dictated this choice.

10. *The Flux of the Quantitative*

One need only consider the nature of quality to realize how poorly it lends itself to the efforts of our minds to connect the changing with the identical, which efforts are the essence of any explanation of reality. For every quality appears to be something complete in itself; not only does the fact that it exists presuppose nothing outside itself, but it is something intensive and thus seems incapable of being combined with or added to anything else. Such an advantage is exactly what characterizes quantity and what distinguishes it from quality: qualities can only be noted, while quantities can be added. When we add quantities, the identity of what we start with seems to be preserved, even though the result is obviously different. In general, by the use of those mental operations we designate by the generic term *calculation*, two quantities quite easily form a third, while two qualities seem to remain separated by a huge gap. This is the argument Aristotle used against his mentor Plato to show that the latter's panmathematicism was incapable of extracting quality from his quantitative world.[7] Aristotle was right, but it is the other side of the argument that interests us here. It shows us, in effect, how a quantitative world, unlike one that retains quality, can be conceived as being in flux. Consequently, from this point of view a quantitative world is analogous to the reality presented to us by sensation and is capable of being used to explain (but only to a certain extent) the incessant flux of this reality. It is probable that the presence of the quantitative element in our perception (although, as we have seen, our immediate sensation is exempt from it) arises, at least in large part, from the unconscious application of this reasoning. We ourself tried to show how, in a very simple case, an element of quantity can be superimposed on the primitive and purely qualitative data of sensation (IR 379 ff.; Eng. 340 ff.).

It seems obvious, in any case, that this is the source of the claim that the quantitative is the essence of reality, its 'substance'. And the preoccupation with measurement that, as Kelvin has rightly observed, characterizes modern physics is only the most elementary manifestation of this same, more or less unconscious, *panmathematicism*.

11. *The Intelligibility of Reality*

Thus, what is behind the predominance of measurement in physical science is the profound conviction that reality is intelligible. It is a conviction that d'Alembert expressed with great clarity. Speaking of physical observations interconnected by deductions, he said:

All of these properties gathered together offer us, properly speaking, only a simple and unique piece of knowledge. If others in larger quantity seem detached to us and form different truths, we owe this sorry advantage to the feebleness of our intelligence.[8]

But, as a matter of fact, it cannot be doubted that all our knowledge is based on this implicit and essential postulate. Auguste Comte denied it, and in this respect positivism has remained faithful to the teachings of its master; as we have often pointed out, however, this position is completely untenable. It is perfectly obvious that our intellect is never content with a simple description of a phenomenon; it always goes further. The knowledge it seeks is not purely external and designed solely to facilitate action, but is an internal knowledge permitting it to penetrate into the true being of things. The world of common sense is only a step in this direction, the initial result of our intellect's constant effort to understand reality. We must, therefore, suppose that the intellect of animals (at least higher animals), who apparently *perceive* things much as man does, is similar to ours in this respect.[9]

Noted physicists reflecting on the foundations of their science have often acknowledged more or less explicitly the essential role of our tendency to seek the intelligible.

Planck, for example, in a résumé devoted to the *principle of least action*, observes that it is important to note "that the strong conviction of the existence of a close relation between natural laws and a higher will has provided the basis for the discovery" of this principle by Leibniz and Maupertuis, as well as their followers (PRB 113; Eng. 76). Analogously, Wien maintains that man's uncompromising drive toward research, the drive that has created the sciences, leads to the supposition that nature is intelligible, that is to say, that there is an agreement between the phenomena of nature and the conclusions of our logical thought. In another passage he observes that the physicist must "presuppose not only the existence of the external world, but also the fact that this world is capable of being understood by us." Wien does not hesitate, moreover, to compare this attitude to Hegel's:

It is only just to recognize that Hegel's assumption that the universe is the product of the thought of a creative mind and that our task is to rethink this thought is basically nothing other than the assumption of the intelligibility of nature.

We shall see below why this comparison is particularly interesting in the context of our present study.

12. *Auguste Comte's Protest*

Thus the predominance of the quantitative is an integral part of what constitutes the very essence of science. This is because we seek to know phenomena only in order to reason about them, and, of the things we know, only those about which we can reason seem really important to us.

Since this is so clearly true, and since the concern for action and prediction is so clearly unable to give rise all by itself to the convictions underlying the predominance of mathematics in modern physics, even Auguste Comte loudly protested Condorcet's opinion that there is no real certainty outside mathematics, declaring that this was a simple prejudice and that, furthermore, chemistry and physiology, "where mathematical analysis plays no part," had subsequently demonstrated it to be an inexcusable prejudice.

Now it is not a prejudice; it is a fact, a fact accepted today by anyone at all imbued with the scientific mentality.

NOTES

[1] Cf. Lucien Poincaré, *La physique moderne* (Paris, 1906), p. 22 [Kelvin, 'Electrical Units of Measurement,' *Popular Lectures and Addresses* (London, 1891), 3:80]. Gaston Moch, in his book on the theory of relativity (*La relativité des phénomènes*, Paris, 1923, p. 34), quite forcefully states the case for this way of explaining the predominance of measurement in modern physics. In support of this opinion, he goes so far as to cite the 'Αεὶ ὀ θεὸς γεομετρεῖ attributed to Plato, even though this is the purest possible expression of a mathematicism resulting from the principle of the intelligibility of the universe.

[2] [See Bibliographic Abbreviations, p. xix above.]

[3] Descartes, *Principles of Philosophy*, Pt. 2, princ. 4 [*The Philosophical Works of Descartes*, trans. E. S. Haldane and G. R. T. Ross (Cambridge: Cambridge University Press, 1931; reprint New York: Dover, 1955), 1:255–256]. [The word "opinions" in Meyerson's reference to Democritus refers to Democritus, Fr. 9 (Diels), Sextus *adv. math.* VII, 135 (as in, e.g., G. S. Kirk and J. E. Raven, *The Presocratic Philosophers*, Cambridge, 1963, p. 422). See IR 320; Eng. 292.]

[4] Max von Laue, *Das Relativitätsprinzip* (Brunswick, 1921), 2:30. General Charles Ernest Vouillemin gives a similar definition: "In physics only the result of measurement is considered to be real" (*Introduction à la théorie d'Einstein: Exposé philosophique élémentaire*, Paris, 1922, p. 10) and speaks of "reality, the daughter of measurement" (p. 196).

[5] Henri Bergson, *Durée et simultanéité* (Paris, 1922), p. 88 [*Duration and Simultaneity*, trans. Leon Jacobson (Indianapolis: Bobbs-Merrill, 1965), p. 65].

[6] Wien, moreover, shows how much Mach's position resembles Goethe's on this point. Because of his strict positivism, Mach wanted to bring physics back to the search for direct relationships between sensations. In so doing he diverges from the position of science, which, as it progresses, tends more and more to eliminate the part played by the subject in observed phenomena (cf. also Wien, WW 124).

One can see how accurate this comparison is by reading the recent book by Petzoldt, who, as we shall see below, claims to be Mach's faithful disciple and, as such, seeks to prove that the theory of relativity conforms entirely to the scheme proposed by Mach. Petzoldt sings the praises of Goethe, particularly insofar as he was repelled by the "grey" system of mechanical science, adding that "we must learn to look though the eyes of the physiologist of sensation and the painter" (SR 70–71) — but does not explain to us, to be sure, how it is that by following this program Goethe arrived at concepts of which physics has never taken the slightest notice.

[7] Aristotle, *Metaphysics*, Bk. 1, Ch. 9, 991^b 21–28, 992^a 10–24.

[8] Jean Le Rond d'Alembert, 'Discours préliminaire des éditeurs,' *Encyclopédie* (Paris, 1751), 1:ix [*Preliminary Discourse to the Encyclopedia of Diderot*, trans. Richard N. Schwab (Indianapolis: Bobbs-Merrill, 1963), p. 29].

[9] Cf. our 'Le sens commun vise-t-il la connaissance?' *Revue de métaphysique et de morale* **30** (1923) 13–21.

[10] Auguste Comte, *Système de politique positive*, (Paris, 1854), Vol. 4, 'Appendice Général,' pp. 123–124 ['General Appendix,' trans. Henry Dix Hutton, *System of Positive Polity* (London, 1877; reprint New York: Burt Franklin, n.d.), 4:580.

CHAPTER 2

REALITY

13. *The Preservation of Reality*

We must, however, guard against going to the opposite extreme from Comte
and claiming that the form of modern science is uniquely, or at least especially, due to the influence of mathematics. In fact, a glance through the
history of physical concepts suffices to show that this form — which is, or at
least was until Einstein, that of a mechanism (in Chapter 19 we shall deal
with the connection between mechanism and relativism) — was already found
in all its essential traits in antiquity, when mathematics properly speaking
played only an insignificant role. The true motivating force was the concern
to preserve the identity of reality discussed in §7.

Indeed, the tendency to mathematicize physics dealt with in Chapter 1
is, if not actually under attack by, at least dominated by this other tendency,
which is truly characteristic of science as distinct from philosophy strictly
speaking and which attempts to preserve the reality of the representation
that theory means to substitute for the common sense representation.

14. *Sensation and the Object*

The concept of an external reality, as presented in the spontaneous ontology
of our perception, is born of our constant effort to explain our sensations.
Because we cannot conceive how they go about changing, we assume that
they depend on some cause more constant than they are, a cause we are
consequently obliged to place outside our consciousness. This *object* is thus,
first of all, a collection of sensations projected outside the self, that is,
hypostasized. But these sensations are not limited to those I actually experience; on the contrary, the latter play a relatively unimportant role compared to the sensations that, largely as a result of my memory, I consider it
possible to experience. An object is not the hypostatization of fleeting visual,
tactile, or olfactory, etc., sensations; it is the hypostatization of the collection
of sensations of all sorts that I remember having experienced in a great
number of circumstances, or that I imagine I ought to experience in a given
circumstance. Moreover, one can readily see that, in the concept of object,

18

these *potential* sensations prevail over our actual sensations to the point that, as Bergson says, "perception ends by being merely an occasion for remembering."[1] Consequently, these sensations, which we could only experience successively and which would obviously be quite ephemeral, are transformed into simultaneous properties of an object persisting in time. When I touch the end of the handle of my umbrella in the dark, I *recognize* it; I have no doubt that it is there in its entirety, with its metal ribs and the silken fabric that covers it, and even that the handle is made of wood, that is to say, has a well-defined internal composition.

15. *The Search for Consistency*

Thus, it is a collection of ephemeral and apparently contradictory sensations that I try to explain in my obstinate search for consistency, for the 'concept of totality' that constantly intervenes in our thought processes, as Höffding showed with his characteristic clarity and depth in a recent work.[2] I endeavor to show how, taking into account the diversity of the circumstances, these sensations can nevertheless result from a persistent and unique reality, from a truth to be found *behind* these appearances. Plato already realized this, showing that when different observers conceive differently the size and shape of one and the same thing, it is still possible, by means of number and measurement, to form a unique concept that explains this diversity.[3] Surely this is also the meaning of Spinoza's well-known apothegm, "Truth [is] a standard both of itself and of falsity."[4] Furthermore, no one could doubt that this is the principle the scientist follows in his research. Höffding, whose exposition we are following here, has shown to what extent the entire work of Galileo in particular seems to proceed from this point of view. When he found himself confronted by an anomaly that was none other than the phenomenon we today call the rings of Saturn, it was because his imperfect telescope made him see, in this object he justifiably supposed to be unique, images he found impossible to reconcile with one another. But Huygens later succeeded in demonstrating what was actually involved and how this reality explained the appearances that had deceived his predecessors.[5]

16. *The Objects Created by Science*

It should be pointed out that the *truth* sought by Galileo and found by Huygens is an *object* whose reality is in all respects analogous to the reality of common sense objects. Although no one before the invention of the

telescope could have had the slightest idea of the rings of Saturn, we no more
doubt their existence today than we do that of any object whatsoever perceived directly by our sense organs. Why should we be surprised by this? Is
it not clear that in the two cases we are concerned with entirely analogous
concepts? For even if we were to admit, as is sometimes assumed, that touch
is the true sense of the real and that we are therefore immediately convinced
of the existence of objects we are able to touch,[6] it is certain that such
objects are rare compared to the sum total of what seems to us to constitute
the real world and which is revealed to us principally by the sense of sight.
Now the sense of sight — as a glance at a mirror is sufficient to convince us —
is subject to countless illusions. Thus, our belief in the reality of the objects
we perceive is, and in the great majority of cases must be, only the result of
a more or less complex process of reasoning; this reality seems necessary to
us in order to explain our sensations. From this point of view, the situation
is exactly like that of the theoretical entities of science, the only difference
being that what causes us to posit the existence of the latter is not simply
the impressions of our sense organs but the impressions of these organs as
they have been refined by the use of instruments.

Common sense objects and theoretical entitities are so much of the same
nature that there is a continual, and quite often imperceptible, transition
between the two classes. Surely, for our ancestors the stars were nothing but
simple luminous points fixed to the celestial vault. They have undoubtedly
remained just that for a large part of humanity today, and one can recall
that even Hegel compared them to eruptions on the skin.[7] For any of our
contemporaries with the slightest degree of sophistication, they are, on the
contrary, immense celestial bodies, whose reality can no more be doubted
than that of the incandescent mass of the sun. Furthermore, in earliest
antiquity the sun itself was often considered to be a purely ephemeral and
immaterial luminous phenomenon.[8] Surely it would be quite impossible to
identify the moment at which these essential transformations occurred. It is
not even necessary that the objects created by science actually be perceived,
with or without the aid of optical instruments, in order to become real;
their existence can simply be inferred, in the same way that an entity that is
initially entirely hypothetical can lator become just as completely real if the
inferences become more numerous. Molecules and atoms were certainly only
theoretical entities from the time of Democritus to the present era, whereas
they have undoubtedly become a part of physical reality since the work of
Gouy, Perrin, and the Braggs. And nothing is more certain than the fact that
we infer their existence by reasoning analogous to that by which common

sense is persuaded of the existence of any object whatsoever, namely because this assumption accounts for a whole series of observed phenomena.[9]

17. *The Attitude of the Philologist*

Moreover, this very characteristic attitude of the human intellect is not seen exclusively in the case of common sense and the physical sciences; it can be seen in connection with *all* research. For a specific example, take the case of a philologist trying to establish the original text from the more-or-less altered readings transmitted to us through different manuscripts. He certainly will not be able to consider his task completed until he can explain how the observed alterations could have crept into the text in question. "The apparently correct reading," Loisy said, in describing how Renan went about studying a Biblical passage, "had to account in some way for the apparently less probable or false readings that were supposed to be derived from it; only after this final test would Renan make a pronouncement."[10]

18. *Reality and Appearance*

The philologist's procedure is undeniably based upon the conviction that there exists one text from which the others are derived; it is only this conviction that allows him to make a choice between the different readings, rejecting some in favor of others. In the same way, common sense does not attempt to include in the object absolutely all the sensations the object seems to be able to give it. Rather it declares some of them real, while others, even though they may actually be present, are judged to be merely apparent. The trees *appear* to be purple in the distance, but they are *really* green.

It may be instructive in more than one respect to consider this example further. First, it clearly shows that, contrary to what is sometimes said, memory is not simply a weakened sensation. In fact, in this case memory is stronger than momentary sensation, to the point that it may make the latter disappear, since until such time as we have learned to attend to the fleeting sensation, we are completely persuaded we see the color green where there actually is the color purple. Secondly, we also see the inadequacy of the theory that the primacy of the sense of touch is what convinces us that the dimensions and shapes of objects are invariable (even though our vision shows them to be constantly changing). For, in the case we are considering, what masks the immediate visual sensation is another visual sensation.

The truth is that this whole process is possible only because we are not dealing with things of the same order, contrary to what one would at first be inclined to believe. Although the color green of the tree was originally only a sensation just like the color purple, common sense reasoning has transformed it into a property of the thing, that is to say, has located it outside our consciousness. Therefore, as is indicated in our statement of the relationship between the two sensations, the green henceforth becomes part of *reality*, permitting it to be substituted for the sensation of purple, which is considered to be mere *appearance*. Thus, the sensation is entirely subordinated to the perception; it seems to be only the sign pointing to the perception, the way leading toward it. The perception appears to be real, so much so that it sometimes succeeds in making us completely forget what served as its point of departure. As Brunschvicg aptly said, "the true nature of common sense is to turn spontaneously toward things" (EH 406).

Furthermore, it is this trait of our perception that makes prestidigitation possible. The clever performer can make gestures that deceive us, making us swear we *saw* something that never happened at all. Here again, it is not a question of a sensation, but of a perception, a snap judgment about an *inferred* reality, and it is this judgment that is responsible for our error.

19. *The Positivistic Point of View*

Thus, as a result of the circumstances just described, the common sense object is already seen to be necessarily detached in large measure from sensation. And the proof is that it continues to exist when I look away and no longer perceive it in any way.

This evolution continues in science. Such an observation is diametrically opposed to the most essential foundations of the positivistic position, as are the observations we have just made concerning the perfect similarity between the objects created by science and those whose existence is posited by a spontaneous act of perception. As a matter of fact, positivism, as we know only too well, would have science abstain from any assumption concerning the true nature of things. It does so by establishing a system of relations without any substratum, which can only result in a search for connections between our pure sensations, a procedure as remote as possible from the one actually used by science. It should be added that Auguste Comte himself, following his powerful scientific instinct and forgetting his own principles, proclaimed on occasion that the "crude but sensible indications of good common sense" constituted "the real and constant point of departure for all sound scientific speculation" (CPP 3:205). It is easy to find similar statements among our contemporaries, even those who seem to have quite orthodox positivistic credentials. Urbain, for example, declares that "science, the fruit of reason, is an extension of common knowledge." [11] The noted chemist was undoubtedly right, and it is completely impossible to practice chemistry in any other way than by following this principle.

20. *Transcendence*

The close relationship between scientific thought and the concept of a thing, considered as independent of sensation, was noted as far back as Lucretius. Malebranche sharpened this thought by pointing out that, insofar as they are subjective phenomena, sensations cannot be directly measured; one cannot be used to measure another. On the contrary, we must first reduce them to causes existing outside ourselves. Bradley has pointed out how difficult it is to form an idea of the laws of physics without recourse to the "transcendent." [12] A contemporary psychologist, Stumpf, has also stressed the fact that

the phenomena of sensation are never what observation uses to establish relationships conforming to laws, which relationships constitute the object and goal of science.

He points out that

regular succession, as well as coexistence ... occurs only in the case of events we place, and must place, *outside* sense phenomena; we are forced to consider them as happening *independently* of consciousness if we are to speak of lawfulness in general or in any way at all. [13]

Thus science is necessarily and essentially realistic, in the philosophical sense of the term. As Marais justly said, "realism ... is the implicit postulate of any science of the external world." [14]

21. *The True Place of Theory in Science*

It follows that hypothesis, or theory based on representations — far from being a parasitical excrescence (as the strict positivist would have it), or even a simple auxiliary means, accepted temporarily only to be rejected later when one has achieved one's true end, which is law (as is assumed by the mitigated positivism that is the current epistemology among today's physicists) — is an integral part of what is most essential in science. For the scientist could not work without basing his thought on a body of assumptions concerning the substratum of phenomena. Hypothesis is indispensable to him, and whatever he professes to believe in this respect, there is always a hypothesis behind his explanations.

Positivism's error obviously arises from the fact that, although science starts out from the common sense conception, it profoundly modifies this position. This transformation, considerable though it may seem, nevertheless takes place, as we have just seen, according to exactly the same principles by which spontaneous perception constructed its world. Therefore, scientific theory ends up with a representation in exactly the same way as does the world of common sense. It follows from this (as we demonstrated in ES 1:22 ff.) that the distinction between a science that explains and one that would abstain from all representation is much less real than is commonly claimed, because even the latter cannot get along without an ontology. The only difference between them is that, while the latter accepts our *customary* ontology, explanatory theories require us to modify it.

Moreover, the progression of a theory, that is to say, the successive changes in our assumptions concerning the nature of things, is itself necessitated by our desire to maintain, or indeed make more perfect, the consistency discussed above, which is constantly being threatened by new observations. We desire a more and more rational reality, and it is evident from what we have said that science seeks to achieve this by continuing the evolution that led reason from pure sensation to the common sense object.

22. *Planck on the Retreat from 'Anthropomorphism'*

Planck has astutely observed that one of the strongest and most constant characteristics of progress in physics is precisely the fact that it is moving further and further from what he calls "anthropomorphic considerations," that is, those involving the person of the observer, or, in other words, what refers to the *self*. In a more recent work, the noted physicist again underlines this essential idea:

It is impossible to deny that the whole of the present-day development of physical knowledge works towards as far-reaching a separation as possible of the phenomena in external Nature from those in human consciousness.

And he adds:

The theorem holds also in physics, that one cannot be happy without belief, at least belief in some sort of reality outside us. . . . A research worker who is not guided in his work by any hypothesis, however prudently and provisionally formed, renounces from the beginning a deep understanding of his own results (PRB 78–79; Eng. 53–54).

Completely analogous opinions are to be found in Wien. It is true that this scientist, under the powerful influence of an environment that has long been the preferred atmosphere for the mind of the epistemologist, expresses himself on occasion somewhat like an orthodox positivist. For example, he welcomes the fact that

theoretical physics has gradually been stripped of all its metaphysical vestiges and has come to recognize that its principal task consists in formulating mathematically expressed laws and deducing their consequences.

But at other times, when he is grappling with the real problems of science, he speaks quite differently. Thus, in the case of electron theory, he points out that this hypothesis shows us that, "in the quest for knowledge, we must remove ourselves more and more from sensible phenomena and the traditional concepts of physics." He gives his full approval to a statement by the philosopher Külpe, according to whom the true criterion for the reality of a concept is its independence from the observing subject, adding:

The physicist cannot seek knowledge, nor in the final analysis can anyone else, without postulating a well-defined external world existing objectively and governed by immutable laws.

In another passage he claims that "the foundation of physical thought can be considered" to be the tendency "to eliminate man's subjective conception and uncover the immutable and independent laws governing the way in which man observes things." [15]

We shall have occasion in a later chapter (§51) to come back to Wien's allusion to the search for "immutable laws." For the moment, let us simply remember his important observation that scientific reality, by its nature as well as by its origin, resembles in all respects the reality offered us by spontaneous common sense perception. At the same time we must note that if, in virtue of the process described by Planck and Wien, the entities created by science are destined to be substituted for those of common sense, they will necessarily be more detached, more independent of the subject, that is to say, more real, than the latter. This is true, for example, in the case of the atoms or electrons that are to replace the material bodies of our spontaneous perception.

NOTES

[1] Henri Bergson, *Matière et mémoire* (Paris, 1903), p. 59 [*Matter and Memory*, trans. Nancy M. Paul and W. Scott Palmer (London: Allen and Unwin, 1911), p. 71].

2 Harald Höffding, *Der Totalitätsbegriff* (Leipzig, 1917), *passim*.

3 Plato, *The Republic*, X, 602C–603A.

4 Spinoza, *Ethics*, Pt. 2, prop. 43, note: *veritas est norma sui et falsi* [*The Chief Works of Spinoza*, trans. R. H. M. Elwes (Bohn Library ed.; reprint New York: Dover, 1951), 2:115].

5 Höffding, *Der Relationsbegriff* (Leipzig, 1922), pp. 94–95.

6 Cf. below, §170 ff., on the role of the sense of touch in the concept of material body.

7 Hegel, *Naturphilosophie, Werke* (Berlin, 1842), Vol. 7, Pt. 1, pp. 92, 462 [*Hegel's Philosophy of Nature*, trans. A. V. Miller (Oxford: Clarendon Press, 1970), pp. 62, 297].

8 Gottlob Frege, 'Uber Sinn und Bedeutung,' *Zeitschrift für Philosophie und Philosophische Kritik* 100 (1892) 25, has rightly said that "the discovery that the rising sun is not new every morning, but always the same, was one of the most fertile astronomical discoveries" ['On Sense and Reference,' *Translations from the Philosophical Writings of Gottlob Frege*, trans. Max Black (Oxford: Blackwell, 1960), p. 56].

9 Henri Poincaré understood perfectly the extent to which our belief in the entities created by science is analogous to the one manifested in common sense. He has even expressed this opinion in a particularly difficult case, that of the existence of the ether (it is well-known that many physicists today seriously doubt that it exists). Indeed, after having said that "this hypothesis is found to be suitable for the explanation of phenomena," he added: "After all, have we any other reason for believing in the existence of material objects? That, too, is only a convenient hypothesis" (*La science et l'hypothèse*, Paris, s. d., p. 245 [*Science and Hypothesis*, trans. W. J. Greenstreet (London, 1905; reprint New York: Dover, 1952), pp. 211–212]. Cf. below, §269, concerning the meaning of the term *convenient*.

10 Alfred Loisy, 'Le cours de Renan au Collège de France,' *Journal de Psychologie* 20 (1923) 327.

11 Georges Urbain, *Les disciplines d'une science* (Paris, 1921), p. 8.

12 Nicolas Malebranche, *De la recherche de la vérité* (Paris, 1721), Eclaircissement 11, 4: 277 ff. [*The Search after Truth* and *Elucidations of the Search after Truth*, trans. Thomas M. Lennon (Columbus: Ohio State University Press, 1980), p. 636 ff.]; F. H. Bradley, *Appearance and Reality* (London, 1893), Ch. 11, p. 123 ff.

13 Stumpf quoted by Ernst Mach, *Die Leitgedanken meiner Naturwissenschaftlichen Erkenntnislehre* (Leipzig, 1919), p. 23.

14 Henri Marais, *Introduction géométrique à l'étude de la relativité* (Paris, 1923), p. 96.

15 Wien, WW 156, 220, 285; cf. 221, 223, and our comments on Wien's comparison between the positions of Mach and Goethe (Ch. 1, n. 6). Adolf Kneser similarly declares that "the quest for a real world cannot be dismissed," and that "the true evolution of science comes about in such a way that sensations are suppressed and fade away into insignificance" (*Mathematik und Natur*, Breslau, 1921, p. 31).

CHAPTER 3

THE SPATIAL

23. *The Agreement between Mathematics and Reality*

If science is dominated by the constant concern to preserve the reality of the substratum of sensation, and if, on the other hand, as we saw in Chapter 1, mathematics has exercised and continues to exercise a considerable and constantly growing influence on the evolution of science, it follows that these two tendencies are capable of being reconciled. And it is easy to see that this is really the case, thanks to the close agreement that exists between our sensation and our reason insofar as it expresses itself through mathematics, an agreement that is without doubt the particular trait most characteristic of both the one and the other. This agreement has often been emphasized by thinkers who have used it — depending on their particular point of view — either to demonstrate the reality of the external world or to establish the dependence of the external world on mental concepts. For example, Sophie Germain, whom, as we know, Auguste Comte considered to be one of his forerunners, wrote:

Can it be doubted that a type of being has an absolute reality when one sees the language of calculation, starting from a single reality it has grasped, give rise to all the realities related to the first by a common nature? If the only thing such relationships had to commend them was the fact that our intellect is able to conceive them, how could it happen that the observation of facts should come, by such a different way, to show a structure outside human thought similar to that whose model man finds within himself?[1]

On the other side, the Hegelian, Rosenkranz, is convinced that "arithmetic, geometry and stereometry show that there exists in nature an idealism of proportions and forms that could in no way be explained by the fact of a pure and simple aggregation of eternal atoms."[2] Great mathematicians, for their part, have sometimes exhibited an almost mystical feeling that a sort of preestablished harmony exists between their most abstract speculations — those most apt to seem purely chimerical to the layman — and reality. K. G. Jacobi, for example, although he had maintained the independence of mathematical thought against Fourier, declaring that its sole object was "the honor of the human mind,"[3] has, on the other hand, in his rather successful distichs parodying Schiller, characterized Le Verrier's great discovery as follows:

27

A disciple, eager for knowledge, approached Archimedes. "Initiate me," he said, "into the divine art that has just rendered such admirable services to the science of the stars by discovering a new planet beyond Uranus." "You call this art divine," said the sage, "and it is. But it was already divine before it explored the Cosmos. What you perceive in the Cosmos is only a reflection of its divinity. For among the gods of Olympus number reigns supreme." [4]

In an academic dissertation, this same mathematician has expressed himself in the following terms:

The natural universe and conscious man are both created by God. The eternal laws of the human mind are the same as those of nature, for if this were not so, the universe would not be intelligible. [4]

Similarly, Hermite has noted the following in his papers:

Just as there exists a world of physical realities, there also exists, if I am not mistaken, a whole world that is the collection of mathematical truths, to which we have access only through our intellect. Both worlds are independent of us; both are divine creations. They seem distinct only because of the inadequacy of our minds. For a more powerful mind they are one and the same thing, and this synthesis is partially revealed in the marvelous correspondence between abstract mathematics on the one hand, and astronomy and all the branches of physics on the other. [5]

Obviously, what is behind all these statements is an acknowledgment of the agreement we have just discussed. This agreement is the only thing that makes possible the *mathematization* of reality (to use Brunschvicg's excellent expression) characterizing modern physics.

24. *The Quantitative in Space*

However, precisely because of the circumstances we have described, the role mathematics plays in physics tends to take a quite specific form. It is easily seen that, even though the mathematics of abstract magnitude, that is, algebra, is regularly used is physics, still, whenever it is a question of representing reality, one resorts exclusively to the mathematics of spatial magnitude, namely geometry. This practice is so much a matter of course that, as the reader may have noticed, when we spoke of the role of mathematics in physics in Chapter 1, we were able to pass more or less without transition from one to the other. The reason for this is, of course, that geometrical properties can be transformed into algebraic properties by means of the physical operation of measurement and thereafter treated according to the rules established for algebraic operations. Conversely, results obtained by strictly geometrical methods, where spatial intuition plays an important role, can be translated into purely algebraic language. The parallelism between algebra and geometry strikes us as soon as we enter the intellectual realm. The very name geometry with which we adorn the science of space expresses this parallelism (since the fact that this science has evolved from the operations of the surveyor depends on the relationship we have just discussed), as does the incomparably more significant fact that geometry seems to us to be an integral part of mathematics. We also know that, through the efforts of mathematicians, particularly Descartes's momentous discovery of analytic geometry and subesquent developments in this field, the parallelism between geometry and algebra has been considerably extended. In the course of this work we shall have occasion to speak of one aspect of this analogy, but also of its limitations. For the time being, let us only note the obvious fact that the concept of physical reality in science as we know it is bound up with the concept of the spatial. "One may justifiably assert," said Weyl, "that space is the form of external material reality" (STM 5). This is, of course, a Kantian formula, but we shall see later why it acquires a particular importance coming from one of the protagonists of the new conceptions. The role of the spatial is, it goes without saying, a direct consequence of the fact that, beginning with the representation of reality spontaneously offered us by perception, but wishing to change it to meet the needs of our reason, we keep only that part of the representation able to be subsumed under the category of quantity. Now this is clearly the spatial. That is why Descartes's statement quoted in §6 reduces reality not only to the quantitative but also to the spatial, and why Locke's primary qualities refer equally to space and the occupation of space.

25. Deduction According to Descartes and According to Hegel

Similarly, the most rash — one might even say, the most extravagant — assumptions about the nature of reality seem to become admissible as soon as this reality is tied to space. Thus a modern physicist reading Descartes's *Principles* will feel no repugnance and even no astonishment (unless it be born of admiration); on the other hand, if we confront him with Hegel's *Philosophy of Nature* he will display total bewilderment, followed by (and these terms are not too strong) disgust and something akin to horror. Nevertheless, nothing is more certain than the fact that Descartes, no less than Hegel, tried to deduce reality from nothingness; he went even further along this paradoxical path than the German philosopher was later to do, since the reality he claims to have deduced is more detailed than Hegel's. Thus he purports to have succeeded in drawing from his premises, by pure reasoning,

the heavens, the stars, an earth, and even on the earth, water, air, fire, the minerals and some other such things, which are the most common and simple of any that exist, and consequently the easiest to know.

Elsewhere he writes: I have "become so rash as to seek the cause of the position of each fixed star,"[6] while for Hegel most of these things clearly come under the category of the 'fortuitous', with which rational deduction need not concern itself. But this is because Hegel wanted to go from thought to nature by means of logical categories; he sought to persuade us that the categories are able, in Seth's words, to "take flesh and blood and walk into the air."[7] Quite on the contrary, the nothingness from which Descartes begins is spatial — since his 'matter' is in fact only space pure and simple — and this is what allows him to retain for it, in spite of everything, some sort of appearance of reality. This is also what keeps his excesses from shocking us too much; we tend to view them rather as somewhat trivial in nature, as simple exaggerations on the part of an exceptionally vigorous mind. In contrast, one is immediately struck by the unreality of the Hegelian concepts of *Sein, Dasein, Fürsichsein*, etc., and even Hegel's strongest supporters have never been comfortable with his transition from logic to the philosophy of nature.

26. *The Corporeity of Geometrical Figures*

Dealing with the same subject in an earlier work (ES 2:203, 227), we referred to the fact, somewhat strange at first but nevertheless incontestable, that geometrical figures seem to be endowed with a mysterious sort of substantiality or *corporeity* that permits them to persist in space, outside our consciousness, even though we have stripped them of any properly material substratum. This explains the attempts that have been made to reduce matter to purely geometrical concepts, and we have shown in our examples how numerous such attempts have been, even after the decline of Cartesianism. As recently as our own era, Henri Poincaré felt obliged to protest against the demands of certain partisans of uncompromising mechanistic theories attempting to reduce everything to a matter "having only purely geometrical qualities," and Duhem, dealing with very different theories, similarly observed that they tended to reduce matter to space.[8]

27. *Explanation by Geometrical Figures*

We also observed, in studying the *modalities* of spatial explanation (ES Ch. 8), how little present-day science differs in this respect from earlier science. It can be seen from a quick glance at the theories that have appeared during the last few years, such as Werner's theory of *complexes* or the theories of J. J. Thomson, Bohr, Bragg, Moseley, etc., on the constitution of molecules and atoms, that in all these cases the essential part of the explanation rests, just as it did for Plato, on properties of geometrical figures.

28. *Geometry and Algebra*

Let us also note in this regard that geometry offers the mind resources that go beyond those available to algebra. Clearly neither LeBel's and Van't Hoff's asymmetric tetrahedron nor Werner's octahedron, to mention only two well-known examples, can be conceived except in space.

What, from a purely logical point of view, are the relationships between spatial or geometrical concepts and those that are purely numerical or algebraic? This is a very important and difficult question for the philosophy of the mathematical sciences. We shall not treat it in depth here, but merely observe that our initial impression of the spatial is that it certainly seems to contain elements that cannot be reduced to number, nor even to concepts derived from number by extension. Aristotle had already realized this; in an attempt to demonstrate, against the mathematicism of his mentor Plato, the inability of the purely quantitative to give rise to anything having to do with quality (cf. §10), he had used precisely the example of geometrical concepts in arguing that number could not account for the specificity of a particular line, surface or solid. It is well-known, furthermore, that Kant was of a similar mind in this matter; in particular, he demonstrated the necessity for a genuine *intuition* of space, by the consideration of symmetrical figures (right and left), since the difference between them can be *given* but cannot be explained in any intelligible manner (*dari, non intellegi*).[9] And we can be persuaded that geometricians today are entirely of the same opinion. "Generally speaking," says Borel, "no operation of an analytical or algebraical kind would enable us to distinguish between a right and a left trihedron. ... The only way to do so would be to *show* them, that is to say, to have recourse to the visual intuition of space, supplemented by the natural distinction which from our childhood we have been in the habit of making between our right hand and our left" (ET 103; Eng. 88). Any geometrician who ever reflects on his work must be clearly aware of this fact. Although he may be persuaded, in Henri Poincaré's words, that "one can do nothing today without analysis," he nevertheless senses "how precious is this thing we call geometrical intuition." [10]

Thus the geometrical undoubtedly harbors an element that is not purely rational in nature, an element of quality. And it is precisely this fact that would seem to account for the advantage referred to above that makes the spatial, unlike the purely algebraic, appear to be endowed with a certain reality, and consequently suitable for representing the real. "Ask your imagination," cries Tyndall, speaking of hypotheses on the nature of light, "if it will accept a vibrating multiple proportion." [11] Lodge, in another treatment of wave theory, makes a similar statement: "Waves must be waves of something." [12] If, however, a geometrical concept had been set in vibration instead of a purely arithmetic multiple proportion, Tyndall would surely have found the theory incomparably more plausible, and Lodge would not have objected to accepting a concept of the same sort as representing the *thing* constituting the wave. At least this is what the examples we have cited authorize us to believe.

29. *Explanation by Motion*

The most essential aspects of the forms of spatial explanation discussed above are very ancient. There is one form, however, that appears to be peculiar to modern science, namely, explanation by motion. Indeed, it could not have occurred to anyone to use this form of explanation until the permanence of motion — its conservation — had been established. This was accomplished only by the law of inertia, which was obviously unknown in antiquity. But this law itself, if examined from the present point of view, is seen to belong to the class of spatial explanations, as we shall demonstrate in the next chapter.

NOTES

[1] Sophie Germain, 'Considérations générales sur l'état des sciences et des lettres aux différentes époques de leur culture,' *Œuvres philosophiques* (Paris, 1878), p. 157.
[2] Johann Karl Friedrich Rosenkranz, *Hegel als deutscher Nationalphilosoph* (Leipzig, 1870), p. 329.
[3] Quoted in Federigo Enriques, 'La critique des principes et son rôle dans le développement des mathématiques,' *Scientia* 12 (1912) 178.
[4] Quoted in Adolf Kneser, *Mathematik und Natur* (Breslau, 1918), pp. 10–12.

[5] Quoted in Gaston Darboux, *Eloges académiques et discours* (Paris, 1912), p. 142. We should note that this statement, although it is later than K. G. Jacobi's statements, is entirely independent of them, since the latter were not published until after Hermite's death. Cf. Kneser, p. 12. Analogously, in a letter to Stieltjes Hermite writes: "I am . . . quite convinced that corresponding to the most abstract speculations of Analysis are realities existing outside ourselves that we shall some day come to know. I even believe that the efforts of geometricians, unbeknownst to them, receive guidance that makes them tend toward such a goal, and the history of science seems to prove that an analytic discovery takes place at the very moment it is needed to make it possible for us to advance in our study of those phenomena of the real world that are accessible to us" (*Correspondance d'Hermite et de Stieltjes* (ed. B. Baillaud and H. Bourget, Paris, 1905, 1:8). There is no doubt that this agreement in mathematics is what d'Alembert had in mind when he spoke of a "simple and unique piece of knowledge" (cf. §11 above). It is completely characteristic, on the other hand, that Petzoldt, whose positivistic bias we have noted (Ch. 1, n. 6), loudly protests against this concept of a preestablished harmony between pure mathematics and physics (which had also been affirmed by Minkowski), declaring that "this is an extremely dangerous belief" (SR 123).

[6] *Discourse*, Pt. 6 [*Philosophical Works of Descartes*, trans. Elizabeth S. Haldane and G. R. T. Ross (Cambridge, England: Cambridge University Press, 1931; reprint New York: Dover, 1955), 1:121] and 'Lettre à Mersenne,' 10 May 1632, *Œuvres de Descartes* (Paris: Ch. Adam et P. Tannery, 1897–1910), 1:250 [*Philosophical Letters of Descartes*, trans. Anthony Kenny (Oxford: Clarendon Press, 1970), p. 23]. Cf. *Principles*, Pt. 4, princ. 199: "And thus by a simple enumeration it may be deduced that there is no phenomenon in nature whose treatment has been omitted in this treatise" [Haldane and Ross, 1:296]. For Hegel, cf. §89 below.

[7] Andrew Seth, *Hegelianism and Personality*, 2nd ed. (Edinburgh, 1897), p. 132.

[8] Henri Poincaré, *Electricité et optique* (Paris, 1901), p. 3; Pierre Duhem, *L'évolution de la mécanique* (Paris, 1902), pp. 177–178.

[9] Kant, *Metaphysical Foundations of Natural Science*, trans. James Ellington (Indianapolis: Bobbs-Merrill, 1970), p. 23 [Preussische Akademie ed., 4:484]. Meyerson cites *Premiers principes métaphysiques de la science de la nature*, trans. Andler and Chavannes (Paris, 1891).

[10] 'Allocation de M. H. Poincaré,' in Gaston Darboux, *Eloges académiques et discours* (Paris, 1912), p. 453.

[11] John Tyndall, *Fragments of Science* (London, 1871), p. 136.

[12] Sir Oliver Lodge, 'The Aether of Space,' *Nature* 79 (1909) 323.

CHAPTER 4

THE PRINCIPLE OF INERTIA

30. *Absolute Motion*

Until the establishment of the principle of inertia by Galileo and Descartes, motion had always been conceived as absolute. This is especially clear in Aristotle, and nothing suggests that the opinions expressed by the atomists on this matter were any different from those of the Stagirite. Moreover, this conception is consistent with man's natural inclination, which is certainly to consider motion as something real in itself. Sir Thomas More, protesting against the Cartesian concept of inertia, writes: "When I am sitting quietly and another man, putting a mile's distance between us, is red with fatigue, it is he who is in motion and I who am at rest."[1] This is a point of view that can easily be found among our contemporaries, even quite well-read people, if their education, as so often happens, has somewhat neglected the physical and mathematical sciences. Furthermore, it is not altogether clear that the position is blameworthy in all respects. Well before Einstein, there was no lack of attempts to explain inertial motion (somewhat in the manner of Aristotle's "environmental reaction") by an action of the medium (IR 116; Eng. 115). And as to the concept of 'general relativity', which at first glance seems so contrary to any absolute element in motion, it has nevertheless been possible to say that "the most remarkable thing about Einstein's hypothesis is that it constitutes a return to absolute time and space."[2]

31. *The Vis Impressa*

But even when our reason is willing to consider *impulsion* as something real, what it initially arrives at is certainly not the concept of a motion continuing endlessly, but rather the concept of a motion that eventually exhausts itself. This is seen in the theories of the *vis impressa* that preceded, and to a certain extent prepared the way for, the Galilean concept.[3]

32. *Motion as a State*

Let us now try to understand a little more clearly what is behind the modification introduced into our way of conceiving motion by the establishment of

34

the principle of inertia. Descartes, seeking to justify his statement of the principle, writes:

Each individual part of matter always continues to remain in the same state unless collision with others forces it to change that state. That is to say, if the part has some size, it will never become smaller unless others divide it; if it is round or square, it will never change that shape without others forcing it to do so; if it is stopped in some place it will never depart from that place unless others chase it away; and if it has once begun to move, it will always continue with an equal force until others stop or retard it.[4]

Here, it is clear, the weight of the argument is borne primarily by the term *state*; the only purpose of the explanation that follows is to persuade the reader that the state of motion is comparable to any other state and must therefore be conserved like any other state. More than a century after Descartes, Euler likewise feels obliged to insist on this conception in combatting our natural inclination (expressed so well by More). "One must not imagine," he says,

that the conservation of the state of a body is limited to bodies remaining in the same place. This does indeed occur when the body is at rest; but when it is moving with the same speed and in the same direction, one also says that it remains in the same state, even though it changes its place at every moment.[5]

33. *Velocity as Substance*

Thus, our affirmation of the principle of inertia is based above all on the spontaneous acceptance with which we are prepared to welcome any proposition having to do with the maintenance or conservation of a concept. The concept at issue here is velocity: velocity considered as a substance in the philosophical sense of the term. We dealt with this aspect of the principle of inertia earlier, in Chapter 3 of *Identity and Reality*, using it as a point of departure for our elucidation of the notion of the *plausible*,[6] so essential for understanding the true nature of the principles of conservation. We showed in particular that both Galileo and Descartes, in proclaiming the perpetuity of rectilinear and uniform motion, supported their claim almost exclusively by considerations based on the sentiment in question. Galileo did so more or less implicitly, merely asserting that "any velocity once imparted to a moving body will be rigidly maintained as long as the external causes of acceleration or retardation are removed, a condition which is found only on horizontal planes," drawing from this the conclusion that "motion along a horizontal plane is perpetual."[7] Descartes's statements are completely clear and explicit, offering as the unique proof of his principle the assertion that

"God is immutable," that He "is not subject to change and always acts in the same way."[8] However, our earlier treatment needs to be completed. Indeed, in the history of science, just as in any history, an event or a change is rarely the result of a single cause. Of course, in tracing an evolutionary process, the historian often finds it desirable to simplify for the sake of clarity. But this must not prevent him from subsequently adding the finishing touches that will bring his work closer to reality, which is always infinitely complex by nature. Thus we shall now consider the situation from the point of view of spatial and temporal relationships, thereby illuminating a side of this evolution that we did mention, but left a bit obscure.

34. The Action of Space

By assuming that motion and rest were so different in nature that sensation itself would necessarily inform us whether we were in the one state or the other (this is the real meaning of More's objection), Peripatetic physics implicitly claimed that space was acting on the body; the body was regarded as being kept (by a force, in the modern sense of the term) in the place it occupied, since it only left it under constraint and stopped as soon as the constraint ceased to act. In the physics of *vis impressa*, the situation was a little more complicated, because a sort of combined action of space and time was allowed; the *impetus* was supposed to exhaust itself in time, in such a way that space could resume its action (considered to be somewhat as the Peripatetics had described it). The principle of inertia, on the contrary, denied any action by either space or time in these circumstances, by affirming that motion, provided that it is rectilinear and uniform (motion qualified as inertial for this very reason), persists through space and time. It can be verified that the concept of the indifference of time and space to motion was frequently invoked by the supporters of the new concepts.

35. The Copernicans

To start with, it must be noted that at the time Galileo and Descartes proclaimed the principle of inertia, faith in the Aristotelian theory of motion was already badly shaken. Indeed, this theory was manifestly irreconcilable with the Copernican claim that the earth moves. For if we are to suppose that the earth, which we instinctively take to be the model of what is motionless and even immutable, is, on the contrary, in perpetual motion, clearly we can do this only by admitting that, contrary to Aristotle, rest and motion cannot be directly distinguished from one another. In other words, we must admit that one cannot identify a motion in relation to space itself, but only the motion of one body in relation to other bodies, given, in the words of Gilbert (who was, of course, a firm adherent of the Copernican ideas), that "the place is a nothing; it does not exist and exerts no force."[9]

Writing at a moment when the special theory of relativity had only just appeared (Einstein's first publications date from 1905) and had as yet attracted little attention outside a very small circle of specialists in mathematical physics, we used the term *principle of relative motion* to describe the proposition Copernicus and his followers implicitly and constantly applied, even though they never formulated it explicitly.[10] It is extremely interesting to note that this proposition is much closer to Einstein's than is the classical statement of the principle of inertia. For what the Copernicans were affirming was precisely the indifference of sensations and phenomena to *any* motion whatsoever. This principle, as they understood it, was quite incapable of serving as the basis for a sound mechanics; it was, moreover, easily refuted by experience, for there is no doubt that we are able to perceive both acceleration and a change in the direction of motion by their immediate effects. This explains, first, why this principle, as we have just said, was never clearly formulated by the Copernicans, although it was constantly implied by their arguments. It also explains why, when they tried to take account of the motion of bodies on the earth, they resorted to all sorts of more or less complicated devices, even rather bizarre ones, as is seen in the case of Kepler (IR 521 ff.; Eng. 461 ff.).[11]

36. *Newton and Kant*

Finally, these same circumstances account for the fact that the idea of a spontaneous continuation of motion could only triumph in the form Galileo and Descartes gave it, namely, by limiting it to rectilinear and uniform motion. Because of this limitation, the principle of inertia was subject to a sort of internal contradiction, since it considered motion as relative in some respects and absolute in others. This is why Newton simply returned to the affirmation of absolute space, justifying it by the famous rotating bucket experiment, which in fact does show that in this case the centrifugal movement of the water is not determined by motion relative to neighboring bodies.

There is no doubt that Newton was completely right, for there is no way to make something absolute out of something relative, even though the absolute can obviously appear to be relative due to our ignorance; therefore, what is absolute in one respect must necessarily be understood as being entirely absolute. However, even after Newton, this idea of relativity of motion by no means disappeared, as we see in certain demonstrations of the principle of inertia where this notion is more or less explicitly invoked. An excellent example of such a demonstration is the one put forth by Kant:

Every motion as object of a possible experience can be viewed at will either as motion of a body in a space that is at rest, or as rest of the body and motion of the space in the opposite direction with equal velocity.

For him this follows from the fact that absolute motion, that is to say, motion not conceived by reference to material bodies placed in space (in Kant's words, *"relating to a non-material space"*), cannot become an object of experience and consequently *"is for us a nothing,"* since absolute space is "in itself nothing and is no object at all."[12]

Thus, Kant is trying to deduce the Cartesian principle of inertia from the implicit concept of the Copernicans; for this reason his assertion about space seems to reproduce Gilbert's. However, we know that in reality we are dealing with two quite different things, given that the history of science teaches us that the Copernicans, reasoning as we have explained, failed to arrive at the principle of inertia, whose origins we must seek elsewhere (notably in the development of the notion of *vis impressa*), and also given that, from the logical point of view, the foundation of the principle of inertia is clearly the one indicated by Descartes himself. Kant's case is, moreover, all the more significant since this philosopher was, as we know, steeped in Newtonian ideas. Hermann Cohen even went so far as to say that Kant had merely reduced Newtonian science to a philosophical system (Cassirer, ER 12; Eng. 355).

To be sure, Kant's attitude depends on a whole body of ideas concerning the nature of space (we shall discuss this in Chapter 18). Still, the fact that his demonstration did not shock anyone and has even been frequently imitated since that time proves that it must be founded on a vague but nevertheless quite vigorous natural inclination. Consequently, it can be admitted that this feeling has, to a certain extent, helped make the concept of inertia acceptable despite its strangeness in other respects. One also realizes that the reappearance of the Copernican concept in Kant, as in those who more or less followed him in attempting to deduce the principle of inertia *a priori*, arises out of the fact that once the Cartesian principle was firmly established in physics, one rightly felt that the evolution it implied was headed, at least partially, toward the absolute indifference of space to motion claimed by the Copernicans. Thus, from a certain point of view, the Einsteinian principle of relativity is merely developing an idea that undoubtedly existed in germ in the principle of Galileo and Descartes, but was even more clearly contained in the implicit principle the Copernicans had habitually used before Galileo and Descartes.

37. *Space and Inertial Motion*

What the principle of inertia actually does is modify our conception of the way a body behaves in space. We were aware that space is indifferent to *displacement*, that no modification of a body can result from the sole fact that it is in a different place. The Cartesian principle teaches us, in addition, that space is equally indifferent to *motion*, provided that this motion is of a particular nature, namely, uniform motion in a single direction. From this point of view, then, space is no longer for us what it was for the ancients: we have transformed our concept of it by removing a power to act that it was supposed to possess.

38. *Aristotle's Explanation and Ours*

This will perhaps become even more evident if we compare this way of explaining the continuation of motion with the one it replaced. How in fact did Aristotle, believing that a body moves only insofar as it is propelled "like a body carried on a chariot," account for the phenomenon of throwing? He explained it by "environmental reaction," by "the action of the air which, being pushed, pushes in its turn."[13] It is thus the air, a material body filling space, that caused this phenomenon; it was supposed to act on the body in somewhat the same way that bombardment by "ultramundane particles" produces the apparent phenomenon of gravity in Le Sage's gravitational theory (to which we shall return later). In our present conception, on the contrary, the material intermediary is completely eliminated, for what causes the continuation of motion is the nature of space itself, which is indifferent to motion in a straight line. In other words, we have transferred the explanation from bodies to space. And it is certain that our explanation, as an explanation, is superior to Aristotle's. It is more compelling because, although it goes against the norms our immediate perception instinctively makes us adopt concerning absolute rest and motion, it is more satisfying to our reason. Even supposing that one could have completely accounted for the motion of throwing with such a theory (Aristotle obviously provided only a sketch), this theory would have been less rational than the conception of Galileo and Descartes.

39. *Two Possible Kinds of Explanation*

This observation is especially worthy of note because it seems open to generalization. To this end, let us consider things from a somewhat more abstract point of view.

The spontaneous metaphysics of our perception shows us, on the one hand, that the causes of our sensation are located in a being placed beyond the sensation and, on the other hand, that this being is divided into discrete parts, into separate *beings* or bodies. However, these bodies are not *entirely* isolated from one another; if they were, they would eternally remain just what they are and no phenomenon would be explicable. They are, on the contrary, held together by a common bond: the space in which they are located. This being granted, how is it to be explained; on what can the explanation be based? It would seem clear that only one kind of relation can be invoked, a relation between the body and something that is supposed to be different from it. There are, in such a case, only two possibilities: either we are dealing with the relations of one body to another — as in the case of impact or gravitational action — or else we are dealing with the relations of a body to space — as for example in all cases involving the motion of a single body in space.

40. *Geometry and Rationality*

These two sorts of explanations, however, are far from being equivalent from the point of view of rationality, that is to say, from the point of view of the satisfaction they afford the understanding. That which treats space and its functions seems to us to conform very closely to pure reason; no one doubts that geometry is, above all, a deductive science. Is it entirely deductive? In the past, many mathematicians firmly believed so, but as science progresses, this opinion seems to be less and less common, and a great majority of the best minds today would certainly be inclined to see it rather as a sort of physical science, that is, a body of knowledge based on observations that can properly be understood as empirical. For example, Henri Poincaré believes that "geometry would not exist if there were no unchanging moving bodies," and Painlevé is of the opinion that the geometrical axioms concerning invariable figures "state in a refined form the properties of the shape of solid bodies." Einstein states more generally that mathematical propositions "rest essentially on induction from experience, but not on logical inferences only."[14] Geometry, then, is a physical science, but of a very particular sort, in which the empirical datum is extremely reduced, whereas the role of deduction, of reasoning, is immeasurably increased.

41. *Impact*

Unlike the relations between bodies and space, the relations between one body and another seem to be very mysterious if one looks deeply enough.

This has been established beyond doubt by Hume, in what would seem to be a definitive manner. As a matter of fact, if there is one case in which the action of one body on another seems to present itself in its simplest form, the form easiest to grasp, it is certainly contact, or impact. We shall come back to this phenomenon in Chapter 19 and attempt to spell out more clearly how we come to believe it to be intelligible. Now, Hume has established that it is in precisely this case that the mind is entirely incapable of using a purely mental operation to predict what will happen, and this conclusion is completely confirmed by examples provided by the history of science. Of course, scientists have not formulated explicit conclusions of the same sort as Hume's; such investigations into the foundations of our knowledge do not fall within their province. Still, by straightforward attempts to solve the problem of impact, they have established more or less experimentally that the ways believed to lead to a solution of the problem do not work. They have tried in many ways. First they tried to represent bodies as composed of particles of a sort of prime matter, that is to say, a matter having only a minimum of properties, or rather having, apart from those that are purely quantitative, only one property: precisely that property it was hoped would be able to explain the action of this matter on another. This property was thought to be either a sort of hardness or a sort of elasticity. But, aside from the fact that the hardness and elesticity necessarily had to be supposed infinite in order to account for the immutability of the laws of physics, it was impossible to understand *how* particles could manifest either of these two qualities. What kind of cement could bind the parts of a particle together so strongly (obviously, since the particle had measurable dimensions, it had to have parts)? And what could explain the astonishing springiness of these particles? It soon became generally recognized that there could be no kind of reasonable solution along these lines. Therefore, with the help of the Newtonian notion of *force*, theories that can be called *dynamic*, of which Boscovich's is the most completely worked out example, came to be added to the conceptions just discussed, which can be included under the category *corpuscular*. For Boscovich the quasi-material corpuscle is replaced by a center of forces, or rather by a single force that acts instantaneously at a distance, but whose nature is modified with distance. This last assumption was indispensable, for it is clearly impossible to explain phenomena by means of a force that is either uniquely attractive or uniquely repulsive, and Boscovich's force, being unique, therefore had to manifest itself sometimes as attractive and at other times as repulsive. This, however, made the whole system particularly difficult to accept. But, even apart from this flagrant difficulty, one wonders how these centers of force, which one is compelled to represent somehow as gigantic spider webs, each one extending over the whole universe, can nevertheless succeed in moving through the universe. And one also asks oneself how these centers, which are nothing but mathematical points, that is to say, empty, can nevertheless possess mass. This explains why Boscovich's theory had little success among physicists, who, even though they admitted the existence of molecular *forces*, attached them instead to a corpuscular particle. But these mixed theories were no more acceptable, because it was just as impossible to understand the relation between a corpuscle and the force or forces emanating from it as it was to understand how a force could set a corpuscle in motion by coming into contact with it. Thus, any property attributed to the particle necessarily appeared inexplicable or occult, and impact was indeed an unintelligible phenomenon, just as Hume had said.

One can demonstrate how general this implicit conviction had become among physicists by pointing out how easily electrical theories of matter won acceptance at the end of the nineteenth century. In electrical theories of matter, the phenomenon of impact, indeed even the mechanical phenomenon in general, was stripped of its status as a fundamental phenomenon in favor of the electrical phenomenon, in which the action is no doubt supposed to take place over very short distances, but in such a way that one does not attempt to make it seem explicable, at least not by reducing it to a mechanism.[15]

It could undoubtedly be pointed out that all these developments of which we have just spoken — Hume's demonstration, the abortive attempts to explain impact, the electrical theory of matter — came quite late in the evolution of human thought. It could also be noted that the general belief that the phenomenon of impact is rational (a belief, moreover, that did not entirely disappear after Hume's demonstration, either among physicists or among philosophers themselves) seems to indicate that within the innermost recesses of our minds we harbor contrary convictions. This is partially correct. It is quite true that we would *like* to believe that reality is rational and that if impact were to be the archetypical phenomenon, we would like this phenomenon to conform to our reason; at the same time, however, we vaguely *feel* that this is not possible. Furthermore, we know that Hume had predecessors in the Middle Ages and, at bottom, the problem he dealt with is nothing other than the "communication of substances" so familiar to scholastic philosophers (ES 2:385). Given what we know about the way the scholastics depended upon classical thinkers, it would seem dangerous to suggest, strictly on the basis of a lack of documentary evidence, that such a train of thought was entirely unknown to the Greeks. Thus, the action of one body on another certainly appears to us to be incomparably less rational than the behavior of a body in space.

For this reason, we make a phenomenon rational if we succeed in taking it from the former class of phenomena and putting it into the latter. As we have seen, this is exactly the change brought about by the establishment of the principle of inertia. And we shall see that the principle of relativity results in a completely analogous change.

NOTES

[1] 'Morus à Descartes,' 5 March 1649, Descartes, *Œuvres* (Paris: Ch. Adam et P. Tannery, 1892–1910), 5:312.

[2] Jean Becquerel, *Le principe de la relativité et la Théorie de la gravitation* (Paris, 1922), p. 283.

[3] IR 119 ff.; Eng. 116 ff. Sir Arthur Eddington, in examining the foundations of the concept of inertia, states that "it is quite natural to think that motion is an impulse which will exhaust itself, and that the body will finally come to a stop" (STG 136).

[4] Descartes, *Le Monde, Œuvres*, 11:83 [*Le Monde, ou Traité de la lumière*, trans. M. S. Mahoney (New York: Abaris Books, 1979), p. 61].

[5] Euler, *Lettre à une princesse d'Allemagne* (Paris, 1772), Lettre 74, 1:322.

[6] [Meyerson uses the term "plausible" to denote statements "intermediary between the *a priori* and the *a posteriori*." Such statements are instances of general statements which are *a priori*, but, insofar as they are *a priori*, indefinite. Only experience can make them definite, "but in this matter experience plays a peculiar role, in the sense that it is not free," since it must conform to the more general *a priori* constraints involved. Such statements are therefore not strictly *a priori* nor merely *a posteriori*. They are, in this technical sense, *plausible* (IR 159–160; Eng. 147–148).]

[7] Galileo, *Discorsi, Œuvres* (Florence, 1842), 13:154 [*Dialogues Concerning Two New Sciences*, trans. Henry Crew and Alfonso de Salvio (New York: Macmillan, 1914; reprint New York: Dover, 1954), p. 215].

[8] Descartes, *Principles of Philosophy*, Pt. 2, princs. 37, 39.

[9] Cassirer, ER 20; Eng. 362 [Meyerson omits the phrase "it does not exist," but indicates no ellipsis].

[10] *Identité et réalité*, 1st ed. (Paris, 1908), p. 417. We did not think it necessary to change this nomenclature in the second edition, which appeared in 1912 (the passage in question can be found there, IR 517 ff.; Eng. 458 ff.), even though we were already familiar with Einstein's ideas, because, since they first appeared only in the form of *special relativity*, there seemed to be no possible confusion between the classical principle and that of modern relativism. We shall see that this is no longer completely true in the case of general relativity.

[11] The present exposition is simply an abridged version of Appendix 3, 'The Copernicans and the Principle of Inertia' (IR 516–527; Eng. 458–465).

[12] Kant, *Metaphysical Foundations of Natural Science*, trans. James Ellington (Indianapolis: Bobbs-Merrill, 1970), pp. 28, 20 [Preussische Akademie ed., 4:487, 481; the italics indicate phrases not found as such in the Ellington translation, but quoted directly from Meyerson, who cites *Premiers principes métaphysiques de la science de la nature*, trans. Andler and Chavannes (Paris, 1891), p. 15].

[13] Aristotle, *Physics*, Bk. 4, Ch. 8, 215[a] 12–19.

[14] Henri Poincaré, 'L'espace et la géométrie,' *Revue de métaphysique et de morale* 3 (1895) 638; Paul Painlevé, 'Mécanique,' in *De la méthode dans les sciences*, 1st series (Paris, 1910), p. 377; Einstein, GE 32.

[15] This is a résumé of our exposition in IR 63–86, 106–111 (Eng. 68–85, 97–102).

CHAPTER 5

RELATIVISM, A THEORY ABOUT REALITY

42. *The Evolution of the Notion of Space*

We must warn the reader that the aspect of the relativistic conceptions we shall consider is somewhat different from the one that generally seems to have caught the attention of philosophers. They have, above all, set out to examine the modifications of our concept of time required by Einstein's theory. This can perhaps be explained in part by the fact that the temporal aspect is predominant in the theory as it first appeared, that is, *special* relativity. We, on the other hand, shall attempt first of all to understand how the theory in its most complete form (that of *general* relativity) transforms the concept of space. It is only after this investigation that we shall treat the change in the notion of time.

43. *Theories Based on Principles and Theories Based on Representations*

In physics, of course, to understand or to explain means to deduce. A deduction embracing a certain number of facts is called a theory, and there can be no doubt that from this point of view the theory of relativity fully deserves its name, since it asserts that once its fundamental principles have been laid down, a whole body of experimental observations follows completely from them. But physical theories are of two sorts: on the one hand, there are theories based on principles, which start from purely lawlike propositions, utilizing (at least apparently) no hypothesis, no supposition about reality, and, on the other hand, there are theories based on representations, in which hypotheses constitute the essential part.

However, let us recall here that, as we noted in Chapter 2 (§21), even when the physicist seems not to be explicitly framing a hypothesis concerning the true nature of things, it does not follow that he is really eliminating hypotheses altogether (an attitude he could not possibly adopt); it simply means that he does not intend to modify in any basic way our preconceived ideas in this area. Let us also add that, since lawfulness and causality (or rationality) are often almost inextricably entangled in science, a principle that seems to be abstract, such as the principle of inertia, can, as we have

seen, have explanatory implications, whereas a theory based on representations, by nature explanatory, can, and indeed generally does, have a lawlike
content, available for calculation and prediction. This is completely obvious
in the case of Einstein's theory.

44. Relativism and Phenomenalism

We often speak of the *principle* of relativity, and this term is accurate, for
what is involved is .a collection of mathematical propositions following
from a single proposition. But behind the principle there is, as we shall see,
a genuine hypothesis about reality that is indispensable to the initial proposition; we are required to modify our conception of reality.

During a discussion at a recent meeting of the Société Française de Philosophie, we observed that this is open to argument. Taking advantage of the
confusion to which the somewhat ambiguous term *relativity* lends itself,
the supporters of the doctrine of Comte and Mach sought to use Einstein's
concepts in support of their cause, proclaiming that the relativity of space
implies the relativity of our knowledge in all realms of ideas and consequently
proves how futile it would be to try to understand the inner nature of things,
as atomic theories in particular claim to do.[1]

45. Einstein's Opinion

Now Einstein himself, on the contrary, has, almost from the beginning of
his work, held a position far removed from this phenomenalism; he has
clearly and resolutely exhibited the realistic tendencies characteristic of any
physicist who understands his task. In 1905, at the same time his first articles
on relativity appeared, he published a work in which, without knowing
Gouy's results or anything about Brownian motion, he calculated the amplitude of molecular motion, and his formulas were later used by Perrin. He
was thus instrumental in the revival of atomism that provoked such virulent
vituperations from Mach — and quite logically so.[2] Furthermore, Einstein has
not changed his opinion since that time. In 1911, at the *Conseil de Physique*
in Brussels, he insisted with great clarity and vigor on the necessity of representing the strange and perplexing phenomena of black body radiation
"in a concrete form," that is to say, by means of a mechanism that might
really be able to explain the experimental data. Finally, in the recent meeting
of the Société Française de Philosophie mentioned above, responding to the
author of the present work, Einstein declared that Mach was "a deplorable

philosopher"˙and that it is "a shortsighted view of science that led him to deny the existence of atoms." [3]

46. *Eddington*

Will it be said that these opinions of the creator of the theory of relativity are perhaps peculiar to him? Let us examine Eddington's exposition, in which he so admirably succeeded in laying bare the profound implications of the theory and explaining them with an uncommon clarity. Eddington, whose complete relativistic orthodoxy has never been challenged as far as we know, declares very near the beginning of his book (which we shall have many occasions to cite here) that he does not share "the view so often expressed that the sole aim of scientific theory is 'economy of thought' " and that he does not abandon "the hope that theory is by slow stages leading us nearer to the truth of things" (STG 29). In his last chapter he comes back to this essential question, asserting that

the physicist, so long as he thinks as a physicist, has a definite belief in a real world outside him. For instance, he believes that atoms and molecules really exist; they are not mere inventions that enable him to grasp certain laws of chemical combination.

Applying these principles to the foundations of the theory he undertook to explain, he wonders whether these concepts are "merely illustrations of the mathematical argument, or illustrations of the actual processes of nature," and he settles the question by asserting that the first of these two possibilities "would give a misleading view of what the theory of relativity has accomplished in science." Thus "the four-dimensional world is no mere illustration; it is the real world of physics, arrived at in the recognized way by which physics has always (rightly or wrongly) sought for reality." He then explains in more detail what he understands this method to be, invoking the procedure common sense uses in constructing a three-dimensional object. He asks us to consider an object that looks sometimes like a figure of Britannia, sometimes like the portrait of a monarch and sometimes like a thin rectangle. We explain these appearances by supposing that we are dealing with something three-dimensional, an English penny: "no reasonable person can doubt that the penny is the corresponding physical reality." Now "he who doubts the reality of the four-dimensional world (for logical, as distinct from experimental, reasons) can only be compared to a man who doubts the reality of the penny, and prefers to regard one of its innumerable appearances as the real object" (STG 180–182).

One can easily see that this reasoning completely fits the pattern we used in Chapter 2 to represent the thought process by which our representation of reality is constituted, both by common sense and by science.

47. *Langevin, Borel, Jean Becquerel, Weyl and Marais*

Texts as revealing as Eddington's could easily be found among other supporters of the doctrine. In the opinion of Langevin, for example, through relativity the physicist "reestablishes contact with reality" (AG 12).

Borel, in setting forth Einstein's theory, states that "the search for truth is the most noble aim of science" and that "there are no degrees between truth and error" (ET 17; Eng. 14), while Jean Becquerel speaks of the "concern for truth" and "the satisfaction the mind experiences in understanding phenomena more thoroughly"[4] through Einstein's theory. Hermann Weyl, for his part, explains that the real problem of physics is "to get an insight into the nature of space, time, and matter so far as they participate in the structure of the external world."[5] He is concerned with "the objective world which physics endeavors to crystallise out of direct experience" and finishes his book with the following affirmation: The relativist

must feel transfused with the conviction that reason is not only a human, a too human makeshift in the struggle for existence, but that, in spite of all disappointments and errors, *it has developed to the point that it can objectively embrace the truth*. Our ears have caught a few of the fundamental chords from that harmony of the spheres of which Pythagoras and Kepler once dreamed.[6]

Finally, Henri Marais's affirmation of realism cited above (§ 20) was made in reference to Einstein's theory. In the preceding sentence he had urged that "if experience confirms the theory, it is natural to regard it [the Einsteinian universe] as true physical space."[7]

48. *Reality as Independent of the Observer*

It is not too difficult, moreover, to see that this attitude, far from being accidental, is actually required by the spirit behind the doctrine. In fact, what the Einsteinian physicist seeks is a way of representing physical phenomena valid not only for an observer on the earth or on the sun, or even for a superobserver placed somewhere in space and assumed to be immobile (this is, of course, more or less Newton's point of view), but for all possible and imaginable observers at once. Obviously he can do this only by separating the phenomena from the observer more completely than science had ever done before, that is, by locating the phenomenon outside him in a more absolute way. In this sense, Eddington is perfectly right to declare that Einstein "has succeeded in separating far more completely than hitherto the share of the observer and the share of external nature in the things we see happen."[8] It is indeed in this form that Einstein himself first presented this theory to the scientific world,[9] insisting on the fact that general relativity furnishes a representation of the phenomenon independent of any system of reference (that is to say, of anything referring to a particular observer), whatever the motion of the system. Since that time proponents of the theory have consistently remained faithful to this way of presenting their ideas. According to Langevin, for example,

just as geometry asserts the existence of a space independent of the particular coordinate systems by means of which its points are located, thereby permitting us to state its laws in an intrinsic form through the introduction of invariant elements (distances, angles, surfaces, volumes, etc.), physics, by means of the principle of relativity, asserts the existence of a universe independent of the reference system that serves to locate events.

In particular, this is "the affirmation of the existence of a reality independent of reference systems moving in relation to each other" (PR 31; cf. PR 54, AG 17).

Similarly, De Donder observes that

general relativity, by not granting a privileged role to any spectator, brings to light what is really inherent in all phenomena; it makes us see what is intrinsic or invariable in phenomena, doing so with the most convenient instrument, the one best adapted to phenomena. This instrument is space-time as defined by Einstein's equations. It is not surprising, therefore, that the method of general relativity allows us to come closer to the realities of nature.[10]

In his excellent presentation of the theory, Born states that it "is first of all a pure product of the tendency toward liberation from the *self* and escape from vague impressions and speculation." Thus he goes on to speak repeatedly, and rightly so, of *objectification*; he considers this evolution toward the objective as directly connected to the evolution toward the relative, that is to say, to what constitutes the very essence of the theory.[11]

This is so incontestable that scientists have sometimes been a bit shocked by the way certain philosophers have interpreted the physical theory of relativity. According to Kneser,

This theory is quite often understood as if it finally made everything conveniently relative, just as if the comfortable doctrine of the philosopher Protagoras, so free of prejudices, had prevailed – the doctrine according to which man, that is to say, his point of view, his system of reference, is considered to be the measure of all things. Now Lorentz and Einstein do just the opposite; they rightly direct their attention toward the things that must not be considered from the relative point of view [*die sich der Relativierung entziehen*]. The principle of relativity is, as a matter of fact, the principle of the nonrelativity of the real; it demands that the reality implied by the observed phenomena of nature remain immutable with respect to possible modifications of viewpoint and system of measurement, that it be, according to the current expression, invariant with respect to the Lorentz transformation.[12]

49. *From Common Sense to Relativism*

As we saw earlier, Eddington considers the concepts of the theory of relativity to be just as real in all respects as the objects of common sense. But he had previously shown that the former come about through a simple extension, a logical development, of the process that creates the latter, namely by the progressive elimination of everything referring only to individual and transitory observation. We picture "a chair as an object in nature – looked at all round, and not from any particular angle or distance." But then we seek to arrive at "a conception of the real world not relative to any particularly circumstanced observer." This is what general relativity offers us, teaching us to eliminate not only the position, but also the motion of the observer, and this is how we finally manage to free phenomena from "that which is peculiar to the limited imagination of the human brain" (STG 30, 36, Preface).

De Donder brings out the point that, in physics, the concept of relativity has passed through three stages, the first being the one characterized by Galileo's formulae, the second that of Einstein's special relativity, and the third, finally, the stage of general relativity. But the Belgian physicist rightly insists on the fact that already in the first stage (that is to say, in our classical mechanics) there was an invariant (namely, the square of the distance), by virtue of which

we eliminate what is *relative* to the different observers in order to attain the *absolute*, here represented by a distance. All observers will study the same geometrical space, and it is in this setting, established once and for all, that physical phenomena will take place (*La Gravifique*, 6).

This remark serves particularly well to bring home to us how closely the thought process followed by the relativists conforms to the eternal canons of the human intellect, which have shaped not only science but also the prescientific world of common sense. Indeed, this world of absolute invariants placed in the eternal setting of space is not only the world of Galilean and Cartesian mechanics, it is also the world of immediate perception. When we open our eyes in the morning, we *see* objects in the "rigid and inert framework" of space (according to the expression used elsewhere by the same author [*La Gravifique*, 12]), and we use this conception, as Plato points out (§15), to explain not only how one and the same object can seem different to us in different circumstances, but also why *several* observers can see it differently. For we hold this world of common sense in common with other humans and undoubtedly also with animals, or at least the higher animals, a circumstance that seems to have greatly contributed to the formation of this faculty. Moreover, the harmony between discordant, and even at first sight contradictory, conceptions is brought about — still according to Plato — by means of number and measurement, that is to say, mathematics. Mathematics is in its most rudimentary form in common sense, but we cannot doubt that it is there, and it is by refining this instrument — in such a way as to enable it to explain new observations as they come to us, to maintain the harmony and reestablish it when it is disturbed — that we succeed in creating theoretical science. However profound the modifications relativism imposes on concepts we consider essential and immutable, however painful the effort required for us to leave the setting of Euclidean space, one cannot doubt that from the present point of view there is a strict continuity between this latest metamorphosis of scientific theories and the phases that preceded it. We can also turn this relationship around, so to speak, and use terms made familiar by relativity to speak of common sense, saying that the latter, just like the former, seeks an invariant, something more stable than fugitive sensation, which can be substituted for it: the visual or tactual chair-sensation can be totally transformed, indeed can even appear and disappear, while the chair-object will remain what it *is*.

50. *Positivistic Declarations*

Let us add that when one finds scientists making positivistic declarations of principle concerning relativity theory, they are usually making a purely platonic profession of faith that plays no role in their reasoning, which reasoning completely fits the pattern we have just laid out. For example, General Vouillemin asserts that Einstein's point of view "shares some of the characteristics of the view Duhem described in the thermodynamicists,

who admit only the direct results of measurements in their science, as opposed to the atomists, who wish to explain the visible by the invisible."[13]
This statement contains a factual error, since we have just seen that Einstein
himself, on the contrary, actively contributed to the renaissance of atomism
that consciously opposed the thermodynamicist conception; furthermore, he
expressly disapproved of Mach's position, which is analogous to Duhem's on
this question. Moreover, General Vouillemin himself declares a few lines
earlier that "the physicist's ideal is to sort out for all categories of phenomena
that on which all men can agree completely," a task the physicist accomplishes precisely by seeking behind the apparent, but contradictory, reality
another reality that, although it is invisible, makes the contradiction disappear. Similarly, Gaston Moch proclaims as the guiding principle of his exposition that it "must be understood that all metaphysics is excluded from it,"
appealing to Le Dantec's views on this point. But he hastens to add that
"banishing metaphysics does not mean that one intends to abstain from
hypotheses," without specifying, as Auguste Comte had very logically done,
that these hypotheses must not in any way deal with the way phenomena are
produced (CPP 2:268, 312). Later on, therefore, Moch unabashedly explains,
just like the physicists we have cited, how the conceptions of observers in
motion with respect to one another can be reconciled by relativism. In so
doing, he contrasts the "subjective nature of the Lorentz contraction" with
other "objective" alterations. It is true that he is careful to add parenthetically
that the latter term merely indicates that the alterations in question are
"verifiable in any system whatsoever"; however, this is obviously a purely
verbal restriction, given that it is just because he considers them to be valid
for any consciousness whatsoever that the physicist places them above all
consciousness, as existing in themselves. Analogously, speaking of events
whose order can be reversed for a particular observer in relativity theory,
Moch adds: "But if the first [of these events] determines the second, the most
that will happen is that this observer will perceive them simultaneously."[14]
It is completely clear that here determination is understood altogether
metaphysically as having to do with absolute reality.

51. *The Metaphysics of Laws*

No doubt both these authors are trying to make the ideas they expound more
consistent with orthodox positivism, by affirming more or less explicitly that
the reality so undeniably sought by relativism is the reality of a law. General
Vouillemin, for example, although he speaks of "discovering a superior

reality that will assume an absolute character for all [families of observers in translation]," making a contrast between measurements that "will retain a mark peculiar to each observer" and an agreement that "will be accomplished only with respect to another reality, reality in space—time," at the same time explains that the postulate of general relativity amounts to "formulating general laws of nature without allowing anything to intervene except elements intrinsic to the problems, elements therefore measured by observers in any sort of relative motion whatsoever."[15] Similarly, Gaston Moch, after having insisted on the necessity of hypotheses, as we noted above, stresses that their particular utility lies in helping make prediction possible (that is to say, in preparing the way for the formulation of laws). He asserts that "the whole problem of relativity" consists in "preserving physical laws" despite the fact that the observer changes.[16] But, as we believe we have already established, this is an entirely illegitimate way to interpret the principles of positivism. In fact, it amounts to affirming the existence of laws in themselves, that is to say, affirming the reality of a world of relationships deprived of substrata and at the same time independent of consciousness. It is obvious that just the opposite is true: the only relationships we are able to know are those in which our consciousness itself constitutes one of the terms, and a rigorously logical positivism would (in accordance with J. S. Mill's stipulation) allow only rules directly governing the sequence of pure sensations, without any recourse to the concept of a thing located outside consciousness (cf. ES 1:6 ff.).

52. The Idealistic Interpretation

Let us be more precise, however. We by no means wish to claim that the philosopher cannot provide an idealistic interpretation of reality, even though he recognizes that relativity theory, like physics as a whole, presupposes the existence of things independent of consciousness. Earlier, speaking of reconciling the conceptions of different observers, we invoked the immortal name of Plato, and that fact alone would seem sufficient to prove that an idealistic conception is not excluded. But this is properly a question for the philosopher, and we hope we shall not be accused of underestimating science if we say that the scientist is ill-prepared for such an undertaking, which requires quite different resources from those he customarily uses. Moreover, the task of the philosopher will be particularly arduous in this case, for the return to an idealism based on sensation will obviously be all the more difficult, given that physical theory has moved away from the conscious self.

The distance separating them is particularly great in relativism, and if the philosopher really intends to take account of this advanced stage of science, he will have to come to grips with a reality that is obviously hard to deal with from this point of view. But whatever solutions he might suggest, it is certain that, in their complexity, they will bear very little resemblance to the simplistic conceptions advanced by some scientists who profess hostility toward all metaphysics. For it is easy to see that these scientists actually adhere to a rudimentary metaphysics; after having affirmed with Auguste Comte that one must abstain from all assumptions as to the true nature of things, they naively use "the crude but sensible indications of good common sense" as their firm foundation (as the founder of positivism himself had also done), without ever realizing that this common sense implies a whole ontology as an essential postulate. Thus, even the theory of a world of laws existing in itself, which we discussed above, actually rests on quite a different foundation from the one officially attributed to it (as we explained earlier in ES 1:30 ff.). One affirms this metaphysics not at all because one would follow the principles of positivism and leave the existence of things out of consideration. One does so for just the opposite reason: because the existence of the world of things seems so certain that one comes to the point of as-suming that even the relationships between things, relationships determined by the human intellect that contemplates them, must nevertheless exist independently of the intellect.

To be completely fair, however, it must be admitted that there could also be, behind Vouillemin's and Moch's statements, something other than the positivism we have been discussing. In Chapter 19, where we deal with the relationships between relativism and mechanism, we shall have occasion to come back to this question in connection with statements in which this other aspect we have just mentioned — which is phenomenalistic in nature — appears a little more clearly.

For the moment, let us simply observe that the scientist will have to beware of the constant temptation to trespass in the philosopher's domain; for every man, including the scientist, "philosophizes as he lives," as Sir William Hamilton justly remarks; "he may philosophize well or ill, but philosophize he must."[17] In fact, the less conscious he is of this constant and essential function of his intellect, the more he philosophizes (and the worse he does it). Vituperations against philosophers do nothing to alter this fact. For, as Edmond Goblot puts it, precisely in the case of relativism "the philosopher is the least metaphysical of men; the scientist is a bit more so; and the ordinary man is hopelessly so."[18] But, by the same token, the philosopher must not become involved in the physicist's deductions, attempting, for example, to impose on him a so-called philosophical conception of relativism, a conception that, as we have observed, bears little correspondence to the views of Einstein and his followers.

53. The Name of the Theory

From this point of view, the name by which the theory is known might seem to be an unfortunate choice. In fact, as we said at the beginning of this chapter, this name is apt to make one think that the new theory takes the existence of reality itself to be relative to something else, in this case, of course, consciousness. Now, such is not the case. The name is explained historically by two facts. In the first place, what Michelson was trying to do in the famous experiments that served as the point of departure for this whole development was to show the consequences of a movement of the earth in relation to the ether, considered to be at rest. The theory explaining why the only observable movements were those relative to other bodies or other movements naturally took the name of relativity.[19] Secondly, relativism, which of course had to begin like any new doctrine by criticizing what had existed before, undertook this work of demolition, as we know, by first attacking the notion of time. It thus endeavored to show that the observer had no means of judging the simultaneity of distant events as the Newtonian theory had postulated that he did. It was deduced from this that the notion of absolute time was deprived of any physical basis and that it was necessary to substitute for it the notion of a *local time* peculiar to the observer and somehow carried around with him as he moved. But that was, after all, only a beginning, for although the relativist makes a point of specifying the role of the observer in this way, he does so in order to provide a representation of the phenomenon valid for any observer whatsoever. Let us add, however, that from a different point of view, the name of the Einsteinian doctrine can be justified quite well, as we shall see in Chapter 24 (§267). But, in any case, one would be misled by appearances if one believed that relativity theory connects the phenomenon any more strongly to the observer, or (to use the philosophical term) to the self.[20] Quite on the contrary, the phenomenon is more detached from the observer, as we have seen, and the reality of relativity theory is certainly an ontological absolute, a veritable being-in-itself, even more absolute and ontological than the things of common sense or pre-Einsteinian physics.

54. *The Reality of Time and Space*

Moreover, it must be admitted that the relativists, in their capacity as scientists, have perhaps not always expressed themselves with absolute precision on this question. Even if we leave aside some general formulas clearly inspired by the desire to conform to this or that philosophical position, we can point to a certain number of passages that seem to contradict the nevertheless quite explicit statements we have cited. These passages are mainly concerned with the theory of time in particular and thus refer to the evolution of which we have just spoken, or else they relate more generally to the fact that time and space, as conceived by common sense, and by pre-Einsteinian physics as well, are no longer considered to be necessary parts of reality. "It is perhaps only the theory of relativity," says Weyl, "which has made it quite clear that the two essences, space and time, entering into our intuition have no place in the world constructed by mathematical physics" (STM 3). But this mathematical construction nevertheless leaves reality intact, and the goal of relativity theory is precisely to inform us about the nature of this reality.

Still, we shall see in Chapter 17 that the opinions expressed by scientists in this area sometimes exhibit real contradictions, and at that time we shall try to identify the profound source of this anomaly. That such an anomaly exists, however, in no way invalidates the conclusion one must reach, given the attitude shared by all the relativists, as the reader has just seen, and given also what might be called the internal logic of their doctrine, which agrees completely in this respect with the general canons of all physical science.

55. *The Vehemence of the Controversy*

Finally, even in the absence of any other proof, if one takes account of the resistance the theory has met — the often very bitter controversy it has stirred up, the outcries from all sides, often by men with no competence to engage in such debates — this alone would be enough to convince us that the object of the quarrel is something quite different from a law or a body of laws. For, Auguste Comte to the contrary, law always seems to us to be outside things, somehow aloof from the concept of reality we cannot live without — unless, of course, it contains elements affecting precisely this concept, forcing us to modify it. That is why physics, chemistry, geology, etc. have never, in the case of simple laws, had the same kind of violent and resounding battles that have always marked the rise and fall of theories. Conversely, if we have indeed grasped that the theory of relativity modifies our representation of reality, the vehemence of the controversy taking place before our eyes will prove to us once again the futility of the positivistic claim that would limit science to law and prediction, demonstrating to what extent positivism is actually ruled by assumptions concerning the nature of things.

56. *The Popularizations*

It could certainly be argued that such theories do not constitute real explanations since they engage only a limited number of minds, even among the relatively well-educated. This objection has some merit. It is indeed true that the reality arrived at by means of higher mathematics is not at all of the same nature as that reached by more accessible means and that in some sense it has less explanatory force. Of course, some confirmed relativists would refuse to admit this. However, we find it difficult to contest. Einstein's theories are mathematical conceptions that, at least in our opinion, cannot really be understood apart from the formalism required by this science. This point of view appears to be confirmed by the almost completely negative results of nearly all attempts to 'popularize' the doctrine, although they have sometimes been made by the most highly qualified scientists (including Einstein himself).[21]

Furthermore, this is not a new observation; the Greeks had already made it, as is shown by the well-known anecdote about the mathematician entrusted with teaching the son of one of the Diadochi. When the king reproached him for making the prince work too hard, the unfortunate pedagogue replied: "Sire, there is no royal road to mathematics." We hope we can be forgiven for harking back to such ancient platitudes, but one need only have followed a few of the debates relativity theory has provoked, and still provokes daily, to see how misunderstood this maxim is, although it is demanded by even the most elementary good sense (as opposed to common sense, to use Brunschvicg's apt distinction) (EH 406). Without going so far as to say, with von Laue, that "those who shout the loudest on both sides [that is, among both the supporters and the detractors of the theory of general relativity] have one thing in common: they understand very little,"[22] we shall only say that objections quite often seem to arise from the misunderstanding we have just pointed out.

Some think they have understood, even though their lack of mathematical education clearly did not prepare them in any way for such an exploit, while others, being more aware of their lack of aptitude, succeed in convincing themselves that the theory itself must contain some inherent defect, or that the minds of those who profess the theory must be afflicted with some peculiar deficiency. Trusting in the famous maxim of Boileau, and believing in its universal application, they wait for some ingenious popularizer to come along to explain the new concepts to them in vulgar language without recourse to mathematical symbols. Now clarity is admittedly extremely valuable wherever it can be achieved – although there again, as we have seen elsewhere when citing the example of Pasteur (ES 2:209), it would be a great mistake to believe that the innovator or the initiator himself could clearly see where he was going – but, as a matter of fact, it cannot be achieved everywhere. For language, it goes without saying, is a creation of common sense; it is permeated with common sense conceptions from one end to the other and is therefore constitutionally incapable of expressing anything that departs from them. Consequently, although language can be used without too much difficulty to sum up the acquired knowledge of the sciences that do not concern the reality presented to us by direct perception – such as the historical or sociological sciences – this project is condemned to failure when confronted with the mathematical or philosophical sciences, whose task consists precisely in attacking this reality and more or less completely transforming the idea we had formed of it.[23]

57. *The Level of Knowledge*

Let us also add that those who least understand the relativistic explanations are the very people who could not seriously doubt the good faith of those who claim to understand, as soon as it becomes evident that the distinction between them obviously rests on a difference in the level of knowledge deemed necessary in order to arrive at an understanding. Everyone is familiar with Scaliger's witty commentary on the Basque language: "The Basques speak among themselves as if they understood one another; as for me, I don't believe a word of it." It goes without saying that the famous scholar only meant to illuminate the strangeness of that unique European idiom, which resembles no other; actually, when we see Basques speaking, we do not doubt for a single instant that they are understanding each other. And, analogously, someone who does not succeed in grasping how gravitation works according to Einstein's theory can nevertheless have no doubt – while hearing the explanations of the initiates, witnessing the myriad discussions kindled by the theories, indeed even while skimming through the most accessible parts of the published works – that the relativists understand: he *sees* them understand.

58. *Mathematics and Philosophy*

Renouvier, while recognizing the similarity between mathematics and philosophy, observed with a hint of regret that the public at large did not seem to put the two sciences on the same footing; a man who has not studied higher mathematics will voluntarily avoid reading a book where he sees symbols he does not understand, while in philosophy "everyone would like to be free to enter at will." It can be seen that this observation, so apt on the whole, does not hold here; the protests we spoke of above bear witness to this: in the case of Einstein's theory people claim the right to leap right in without any mathematical preparation. And the reasons for this apparent anomaly are not difficult to recognize. It is quite simply because in this case we are dealing not with mere calculation for the sake of prediction, but with assumptions concerning the nature of things.

Thus, the reason the public does not treat Einstein's theory with the reserve it customarily exhibits toward higher mathematics is that it considers, in Borel's words, that "in spite of what certain scientists are saying, the public feels that there is something here which is of interest to every cultured man" (ET 2; Eng. 2). It is not simply a question of mathematics in this case; the mathematics only serves as the basis for a genuine physical theory, which is to say, as we have explained many times, an ontology. In this sense, relativity theory does indeed belong to philosophy, and thus one could hardly be surprised that the public shows its impatience upon being told it cannot understand a conception that is supposed to bring about a profound modification in its idea of reality. Here, alas, the need does not suffice to create the organ.

59. *The General Public and the Elite*

Nevertheless, objections to relativistic reality arising out of such considerations should not be taken too seriously. Indeed, nothing is more certain than the fact that in contemporary physics a great many theoretical explanations, even some of the most essential ones, actually remain completely inaccessible to anyone ignorant of infinitesimal calculus. Thus, someone who has studied what is customarily called "experimental physics" believes he understands that motion can be transformed into heat and *vice versa* because energy must be conserved. But in believing this, he is deluding himself — or rather he is deluded. He thinks he knows what energy is, because, by a sort of sleight of hand (if this expression is not too disrespectful), 'experimental physics' has provided him with a verbal definition of this concept, telling him, for example, that energy is the capacity to produce an effect. Now this definition is manifestly false, for it expresses a totally different concept, that of Gouy's "available energy." As for 'energy' pure and simple, it is impossible to give a verbal definition of it, for the quite simple reason that it is nothing more than an integral. Therefore, on this very important point — since it is a question of something that is conserved and consequently a case in which the sequence of phenomena seems to us to conform to the requirements of reason — anyone ignorant of higher mathematics cannot claim to have *understood*. There is, furthermore, little need to point out that the situation is not much different even outside the domain of mathematical physics proper. Does anyone believe, for example, that Hegel, when he 'constructed' reality in its entirety using concepts that can be reached only after a quite difficult process of abstraction (to say the very least), was under the illusion that this explication could be made accessible to the ordinary mortal? On the contrary, he was obviously thinking only of an elite. And we believe that Einstein, for his part, would be mistaken to think of a different audience.

NOTES

[1] Petzoldt's book, *Die Stellung der Relativitätstheorie in der geistigen Entwicklung der Menschheit* [SR], which we have already had occasion to mention (§ §8, 23), presents this point of view with great rigor and force.
[2] It can be noted in this context that Mach consistently and explicitly protested against the kinetic theory with which Boltzmann and Maxwell, making use of the calculus of probabilities, explained Carnot's principle, although this theory is undoubtedly one of the cornerstones of contemporary physics. Mach renewed his protest in 1919 (*Die Leitgedanken meiner naturwissenschaftlichen Erkenntnislehre*, Leipzig, 1919, p. 9). We know that Auguste Comte, anticipating in some sense, had characterized any application of the calculus of probabilities to physics as "either childish or sophistical." This is certainly yet another proof of the spiritual kinship between Mach and the founder of positivism. It is hard to say what basis Petzoldt has for claiming that Einstein had, from this point of view, "grown up in the intellectual atmosphere primarily created by Mach's *Mechanics*" (SR 92).
[3] Cf. *Bulletin de la Société française de philosophie* 22 (1922) 112 (meeting of 6 April 1922), [Einstein's remarks are translated in *Nature* 112 (1923) 253; we have translated directly from Meyerson's text, which does not agree exactly with either the official version in the *Bulletin* or the translation in *Nature*.] Likewise, during a discussion that took place in Prague in January 1921, Einstein is said to have explicitly rejected the idea of an analogy between his ideas and Mach's positivism (cf. Aloys Müller, *Die philosophischen Probleme der Einstein'schen Relativitätstheorie*, Brunswick, 1922, p. 14).
[4] Jean Becquerel, *Le principe de la relativité et la Théorie de la gravitation* (Paris, 1922), p. 2.

5 We have translated from the original (Hermann Weyl, *Raum, Zeit, Materie*, 4th ed.,
Berlin, 1921, p. 125), since the passage beginning "so far as," which seems essential
to us, was omitted in the translation (*Temps, espace, matière*, trans. G. Juvet and R.
Leroy, Paris, 1922, p. 119). [This passage is included in the Henry L. Brose translation
(STM) quoted here (p.137).]

6 *Temps, espace, matière*, pp. 189, 275 [STM 217, 312; the italicized phrase is a direct
translation of the French, which departs significantly from the Brose translation at
this point; the latter reads: "it is yet able to follow the intelligence which has planned
the world, and that the consciousness of each one of us is the center at which the One
Light and Life of Truth comprehends itself in Phenomena"].

7 Henri Marais, *Introduction géométrique à l'étude de la relativité* (Paris, 1923), p. 96.
In spite of Marais's background, his book is purely scientific. But we can also cite a
philosophical work that makes the same point. Thus, for Hans Reichenbach, "when
he [Einstein] says that space is curved, this is true in the same sense that one says
that the earth is a sphere. If one wishes to avoid Einstein's geometry in order to seek
refuge in [philosophical] relativism, one must give up the spherical shape of the earth
and, in general, every geometrical shape attributed to real objects" ('La signification
philosophique de la théorie de la relativité,' trans. Léon Bloch, *Revue philosophique*
94 [July – Aug. 1922] 40 [all bracketed insertions are Meyerson's]).

8 STG Preface [the bracketed insertion is Meyerson's]. Cassirer very soundly characterizes the contribution of relativity theory to progress in physics as Planck conceives
it, saying that, by virtue of Einstein's ideas, "anthropomorphism . . . is here again forced
a step further back" (ER 42; Eng. 381 – 382). It should be remembered in this connection that Planck, who is one of the strongest supporters of the theory of relativity,
vigorously opposed Mach's theory ('Die Einheit des physikalischen Weltbildes,' *Physikalische Zeitschrift* (Leipzig, 1909), *passim*; cf. Mach's critique, *Die Leitgedanken*,
p. 7 ff.

9 Einstein, 'Zur allgemeinen Relativitätstheorie,' *Sitzungsberichte der Königlich Preussischen Akademie der Wissenschaften* (4 Nov. 1915), p. 800.

10 Théophile De Donder, *La gravifique einsteinienne* (Paris, 1921), p. 12.

11 Max Born, *La théorie de la relativité d'Einstein*, trans. Finkelstein and Verdier
(Paris, 1923), p. xi [this is a translation of *Die Relativitätstheorie Einsteins* (Berlin,
1920)]: "The result of Einstein's theory is thus the relativization and the objectification of the notions of Space and Time." Again, p. 317: "the most complete objectification and relativization imaginable." [These passages are not to be found in the
English version, *Einstein's Theory of Relativity* (New York: Dover, 1962), which is
heavily revised; see its preface, pp. iii – iv. We have translated directly from the French
version.]

12 Adolf Kneser, *Mathematik und Natur* (Breslau, 1918) [Meyerson has inserted the
German; he gives no page number]. Petzoldt, seeking to demonstrate that relativism
conforms completely to Mach's positivism, actually sings the praises of Protagoras
and goes so far as to say that "it is humanity's misfortune that the sound doctrine
of this philosopher could not prevail against the doctrines of Plato and Aristotle; we
would have been spared the Middle Ages" (SR 15 –17).

13 General Charles Ernest Vouillemin, *Introduction à la théorie d'Einstein* (Paris, 1922),
p. 11.

14 Gaston Moch, *La relativité des phénomènes* (Paris, 1922, pp. 23, 24, 135, 142).

62   CHAPTER 5

[Moch is mistaken. If two events are causally related they will not be simultaneous in any frame of reference.]

15 Vouillemin, *Introduction*, pp. 174, 193, 194; cf. p. 197 [the bracketed insertion is Meyerson's].

16 Moch, *La relativité*, pp. 24 ff., 59 [the bracketed insertion is Meyerson's] ; cf. his statement of the principle of general relativity, p. 226: "the laws of nature are the same for observers in any state of motion whatsoever relative to one another." Petzoldt (SR 120) also maintains that the invariance sought by relativity is an invariance of laws. Cf. above, §22.

17 Sir William Hamilton, *Lectures on Metaphysics and Logic*, 6th ed. (Edinburgh, 1877), 1:65. But did not Kant already point out that the metaphysical illusion is so "inseparable from human reason" that "even after its deceptiveness has been exposed [it] will not cease to play tricks with reason and continually entrap it into momentary aberrations ever and again calling for correction" (*Critique of Pure Reason*, trans. Norman Kemp Smith, B354–355)?

18 Edmond Goblot, 'Einstein et la métaphysique,' *Revue philosophique* 94 (July– August 1922) 136.

19 The substitution of the generalized theory for special relativity (which Einstein set forth initially) is a move in the same direction. Indeed, the motion of the water in the turning bucket, which Newton considered as taking place in absolute space (§36), thereby becomes relative to existing masses. It is Mach, as we know, who first suggested this, and in this respect he is indeed the spiritual ancestor of Einstein (but not with respect to the principles underlying his concept of science; cf. §45).

20 This is the basic thesis Petzoldt develops in SR (cf. note 1 above).

21 This would certainly seem to be the opinion of Eddington, who, even while affirming that he wished to give an account of Einstein's work "without introducing anything very technical in the way of mathematics," nevertheless allows that if one wishes to "abstract from the phenomena . . . that which is peculiar to the limited imagination of the human brain" (which the relativists acknowledge to be their ultimate goal), one can achieve this only by means of mathematical symbolism (STG Preface). Einstein himself, in one of his first expositions (written before his theory had achieved its definitive form) wrote that the road leading to the general theory of relativity "can be described only very imperfectly by the use of words" ('Zum Relativitäts-problem,' *Scientia* 15 [1914] 347). Borel, from the beginning of his brilliant book on *Space and Time* (ET 1; Eng. ix), realizes that a "didactic account" of Einstein's theories "requires the use of the formulae of mathematical physics" and declares that he will be satisfied with a "sort of general survey of the theories." Gustav Mie observes that the theory of relativity is essentially a mathematical theory, adding: "One truly grasps its thought only if one plunges into the mathematical work itself" (*La théorie einsteinienne de la gravitation: Essai de vulgarisation de la théorie*, trans. Rossignol, Paris, 1922, p. 104). Born, whose work is a quite remarkable attempt to explain at least certain aspects of relativity without recourse to any formula of higher mathematics, is nevertheless obliged to admit toward the end of his book that without the aid of methods borrowed from mathematics "a more profound understanding of the general theory of relativity is impossible" (*La théorie de la relativité*, p. 297; cf. p. 282: "the mathematically uninitiated will undoubtedly not be able to find his way through all these expressions" – the German text is still more explicit: "wird sich unter diesen Ausführungen nicht viel denken können," literally: "will

not be able to get much of an idea from these expositions" – *Die Relativitätstheorie Einsteins*, Berlin, 1920, p. 200). Wien judges that the theory "in its essential parts cannot be set out in such a way as to be understood by everyone" and does not believe that one "can hope for any useful effect whatever as a result of the many lectures and articles designed to introduce the theory to those who have not received a scientific education." He concludes by declaring that the famous inscription, "Entry forbidden to all non-mathematicians," engraved above the entrance to Plato's Academy, is even more valid for relativity theory (WW 264, 286).

22 Max von Laue, *Die Relativitätstheorie* (Brunswick, 1921), 2:5.

23 "I can see why the Pyrrhonian philosophers cannot express their general conception in any manner of speaking; for they would need a new language" (Montaigne, 'Apologie de Raimond Sebond,' *Essais*, Paris: Flammarion, s.d., 2:249 ['Apology for Raymond Sebond,' *The Complete Essays of Montaigne*, trans. Donald M. Frame (Stanford: Stanford University Press, 1958), p. 392]).

CHAPTER 6

GRAVITATION

60. *The Mystery of Newtonian Gravitation*

For our present purpose, we shall merely note that in the extended theory of relativity, gravitation, just like inertia, depends on the essential nature of space and that the particular property of space that accounts for gravitation, namely its *curvature*, is clearly mathematical.

Perhaps the implications of this observation will become a bit clearer if we very briefly review the status of gravitation before the introduction of the new theories. In the twelfth century, at the height of the Middle Ages, when gravitation was, of course, considered to be a purely terrestrial phenomenon, John of Salisbury declared the fall of bodies to be a *true* phenomenon but not one that could be understood as *necessary*, using this example to establish (contrary to the then prevailing doctrine) the possibility of a science of that which simply is; we would call this a science dealing exclusively with laws, or a positivistic science.[1] With the formulation of Newton's law, the problem acquired a new urgency, and from then on we find scientists treating gravitational action as a mystery that needs to be explained. It is on such grounds that Leibniz opposed the Newtonian theory, claiming that it was an *idle* hypothesis and accusing it of destroying "not only our philosophy which seeks reasons but also the divine wisdom which provides them."[2] Although there were many philosophers after him who accepted the new notion and even made it the basis for a general conception of reality (IR 80–81; Eng. 80–81), physicists continued to resist it. Their state of mind manifested itself, moreover, in the constant outpouring of theories, often by scientists of great renown, attempting to account for the mystery. There were reputedly more than two hundred of them, a few of whom we have enumerated elsewhere (IR 79; Eng. 80). The most widely accepted theory was that of Le Sage. It is extremely significant in the present context to note that highly competent physicists devoted such time and effort to developing this theory in spite of the truly extravagant assumptions it required. One can only conclude that they were thoroughly convinced that this was a mystery science had to solve at any cost. Among those who held this opinion in the nineteenth century was the illustrious Maxwell, who declared that the mere promise that a theory could lead to an explanation of gravitation would suffice to make men of science devote the rest of their lives to it.[3] Among our contemporaries, Borel called Newton's principle "quite absurd in its classical form" and went on to say that anyone who succeeded in explaining his law (by statistics, for example) would be admired as much as Newton himself.[4] Similarly, Brillouin noted that it was "every physicist's dream, too often unrealized," finally to see "universal gravitation emerge from its magnificent isolation."[5]

61. *The Relativistic Solution*

Einstein's solution is essentially to make gravitation and inertia interdependent, to blend them together, as it were. It had long been known that the same mass was involved in both kinds of phenomena, but this was a bare statement of fact. Newton had believed that this needed to be verified experimentally by comparing pendulums made of different materials. Quite recently, after the discovery of radioactive bodies, which, of course, behave differently in many respects from all other known bodies, it again seemed necessary to demonstrate experimentally that the rule was equally valid for them. For Einstein, on the contrary, this question no longer arises, and thus something new has been achieved.[6]

As a result, the problem that had worried earlier physicists so much, as we have just seen, no longer exists for a follower of Einstein. From the point of view of relativity, Weyl is quite right to speak of "a solution . . . of the problem of gravitation" (STM 227),[7] not only because *instantaneous* action at a distance is henceforth replaced by action operating at the speed of light – it had been noted much earlier that there were inadmissible presuppositions in Laplace's famous calculation attributing to gravitation a velocity fifty or even one hundred million times greater than that of light[8] – but also because the *nature* of this action is determined, so that there is no longer any need to search for an intermediate cause such as Le Sage's "ultramundane corpuscles." This is undoubtedly the most remarkable achievement of the theory, as Einstein himself states in one of his most recent publications:

The possibility of explaining the numerical equality of inertia and gravitation by the unity of their nature gives to the general theory of relativity, according to my conviction, such a superiority over the conceptions of classical mechanics, that all the difficulties encountered in development must be considered as small in comparison (VVR 38; Eng. 58).

62. *The Spatial Nature of the Theory*

Now, the Einsteinian physicist obviously understands things spatially. Actually, the identification of inertial with gravitational mass we have just discussed is not the only identification involved. In relativity there is also fusion of the strictly mathematical properties of space and its physical properties, which were previously thought to belong to clearly distinct domains. Thus inertia was obviously considered a property of physical space, space containing bodies. *A priori* demonstrations such as Kant's (§36 above) could cast doubt on this to a certain extent, since they apparently concern only space in general, but upon more careful examination one becomes convinced that such demonstrations actually presuppose the presence of material bodies in order to indicate both direction and speed. Moreover, Kant's deduction, as we have seen, expressly stipulates that all motion must be related to material bodies placed in space. In giving the principle of inertia its definitive form, Newton made this content explicit; there cannot be the slightest doubt that *absolute* space, as he understands it, is physical space. The confusion between physical and mathematical space — it goes without saying that the term 'confusion' is used here not in a pejorative [but in an etymological] sense — is a peculiarity of the recent theories and clearly distinguishes them from their predecessors. "Gravitation," Langevin tells us, "is understood to be an aspect of geometry." Henceforth, "geometry and physics form a single whole," but this whole is "a geometry of a higher order," that is, it has the form of a science of space, "gravitation being but one aspect of this geometry." Such a geometry consequently "seems destined to absorb the whole of physics." [9]

63. *The Analogy with Previous Theories*

Thus — and this is essential — the relativistic explanation is, above all, a geometrical explanation. Borel even seems to think that this fact constitutes an essential distinction between this theory and the ones that preceded it; indeed, he predicts that, whereas the nineteenth century was the century of mechanistic explanations, the "twentieth century will, perhaps, be the century of geometrical explanations" (ET 208; Eng. 175). Strictly speaking, this statement is not entirely accurate. Actually, as we saw in Chapter 3, any true scientific explanation is, and at bottom can only be, a spatial explanation; therefore it must necessarily involve geometry. Considered, then, from this point of view, relativity is not quite as revolutionary as it might seem at first glance. It is merely continuing the work of providing spatial explanations, while adding a new method to those that physics had previously used.

However, Borel's statement is not without merit, as we shall see below (§81). For the present, let us merely note that if the relativists do understand, it is because for them space is no longer exactly as spontaneous common sense perception presents it, since it henceforth exhibits a property — *curvature* — that such perception does not reveal in any way.

64. *Projectile Motion and Gravitation*

If we compare the way the concept of gravitation changes as a result of Einstein's explanation with the change brought about by the introduction of the law of inertia, we note a striking analogy. This analogy is particularly useful in helping us grasp to some extent, if not how an Einsteinian physicist actually understands gravitation, at least how it is possible for him to arrive at an understanding. Indeed, if we at first found it surprising and unbelievable that an Einsteinian physicist should understand gravitation as he does, it is surely because we imagined gravitation to be a kind of force. For the same reason it seemed completely plausible to accept an explanation like Le Sage's, which replaced Newtonian action at a distance with the impact of invisible particles. But let us reconsider these phenomena we attribute to the mutual attraction of bodies, this time in terms of pure motion or displacement: what we really know about a body supposedly attracted by another — what I see, for example, when I remove the support holding a body up — is that it moves, even though nothing seems to be pushing it.[10] Now, as we said, when Aristotle observed that a hand-thrown object continued to move even after the hand had ceased to touch it, he found this phenomenon to be mysterious in exactly the same sense that gravitation was a mystery before general relativity. This is why the theory he imagined bears so strong a resemblance to Le Sage's theory of gravitation; like Le Sage, he evokes the impact of invisible material or quasi-material particles to explain a visible movement. But Galileo and Descartes make all explanation of this sort superfluous, by simply replacing it with the relationship of the body to space: the body changes place because space possesses the peculiar property of being indifferent to movement so long as it is uniform and rectilinear. Therefore it is not so paradoxical that Einstein, by modifying the properties of space — the relationship of the body to space — could in this way explain the movement that Le Sage attributed to the intervention of a material agent.

NOTES

[1] Joannis Sarisberiensis, *Polycraticus*, Bk. 2, Ch. 21, *Opera Omnia, Patrologie Migne*, 2nd series (Paris, 1855), 199:447: *Scio equidem lapidem vel sagittam, quam in nubes jaculatus sum, exigente natura recasurum in terram, in quam feruntur omnia nutu suo pondera, nec tamen simpliciter recidere in terram, aut quia novi recidere necesse est. Potest enim recidere et non recidere. Alterum tamen, etsi non necessario, verum tamen est. . . . Ceterum etsi non esse possit, nihil impedit esse scientiam, quae non necessariorum tantum, sed quorumlibet existentium est, nisi forte et tu cum stoicis existentia censeas necessariis comparanda.*

[2] Leibniz, *Nouveaux essais, Opera*, ed. Erdmann (Berlin, 1840), p. 203 [*New Essays on Human Understanding*, trans. Peter Remnant and Jonathan Bennett (Cambridge: University Press, 1981), p. 66 (Berlin: Akademie-Verlag, 1962 pagination)].

[3] Clark Maxwell, *Scientific Papers* (Cambridge, 1890) 2 115 ff., 311.

[4] Emile Borel, *Le hasard*, 2nd ed. (Paris, 1914), pp. 3, 300. In his more recent book Borel further explains that "it is because it has effected the entry of universal gravitation into a more general conception of the world . . . that Einstein's theory of general relativity has been received with such admiration and passionate curiosity in scientific circles all over the world" (ET 40; Eng. 33).

[5] Marcel Brillouin, 'Propos sceptiques au sujet du principe de relativité,' *Scientia* 13 (1913) 23. Cf. analogous statements made somewhat earlier by Sir Oliver Lodge, 'The Aether of Space,' *Nature* 79 (1909) 323, and Walther Ritz, 'Die Gravitation,' *Scientia* 5 (1909) 255.

[6] At a time when the general theory of relativity still existed only in some kind of nascent state for Einstein, he indicated how important Eötvös's experiments were for the theory. In these experiments, Eötvös demonstrated with a high degree of exactitude the identity of inertial and gravitational mass, "whose definitions are logically independent of one another" ('Zum Relativitäts-problem,' *Scientia* 15 [1914] 342). Weyl also pointed out the significance of this result (STM 225).

[7] Edmond Bauer, in his excellent little book *La théorie de la relativité* (Paris, 1922, p. 63) makes this aspect of the theory quite clear: "Thus disappears from science the notion of the force of gravitation, of universal attraction – a mysterious property of matter that seemed to make itself felt instantaneously at a distance with no conceivable mechanism."

[8] Cf. IR 80 (Eng. 80). Cf. also, however, the reservations expressed below, §192.

[9] Paul Langevin, AG 19, 22, and preface to Eddington, *Espace, temps et gravitation*, trans. Rossignol (Paris, 1920), p. ii. In order to justify this complete fusion of physics and geometry, which is one of the most essential characteristics of the new doctrine, the relativists take pains to emphasize that "it is only the whole composed of geometry and physics that may be tested empirically," given that the observations themselves constantly involve assumptions such as the one equating the path followed by light rays with a straight line (Weyl, STM 93). Thus, "this seals the doom of the idea that a geometry may exist independently of physics," and "the metrical . . . field is related to the material content filling the world" (STM 220), while, on the other hand, "gravitation is a mode of expression of the metrical field" (STM 226), so that "Geometry, Mechanics, and Physics form an inseparable theoretical whole," which must be conceived *en bloc* (STM 67). In a more recent work this physicist declares, moreover, that the metrical

structure is not rigidly given *a priori*, "but constitutes a *field* describing *the state* of physical reality causally dependent on the state of matter." He adds this picturesque image: "Like the snail, matter itself constructs and forms its own home" (*Mathematische Analyse des Raumproblems*, Berlin, 1923, p. 44).

[10] This is the meaning of the famous thought experiment based on the image of a *suspended chest*. First set forth by Einstein himself (cf. his RSG 66), it has since been used in a great number of expositions of general relativity, particularly those that are not purely mathematical. This line of reasoning does serve especially well to demonstrate that it is impossible to distinguish a gravitational field from a field of accelerated motion. It is only after we have understood this, after we have assimilated the action of gravity to a pure and simple motion, that an explanation of gravitation in terms of the properties of space becomes possible.

CHAPTER 7

TIME

65. *Minkowski's View*

If we now turn back to an aspect of the new theory we had temporarily set aside, namely, the manner in which it tends to modify the ordinary concept of time, we shall readily recognize that here again it is properly a question of spatialization. "Henceforth," Minkowski says, in setting forth the fundamentals of his conception, "space by itself, and time by itself, are doomed to fade away into mere shadows, and only a kind of union of the two will preserve an independent reality." A little later in the same fundamental exposition of his theory he repeats that "space and time are to fade away into shadows, and only a world in itself will subsist." [1]

66. *The Views of Langevin and Wien*

It could certainly be pointed out that what we have here is only the personal opinion of this talented mathematician and that, despite the important role he played in creating the theory (it has been said that the theory of general relativity would have been inconceivable without his formulas), [2] advocates of the theory frequently profess less extreme opinions. Langevin, for example, actually writes: "We are certainly not saying that time is a fourth dimension; that would make no sense" (AG 6). Likewise, Wien declares that "although the relationship between space and time as revealed by relativity theory is very important, one must always bear in mind that we are dealing with a purely formal connection in this case, as follows from the fact that it is not time itself that plays this role, but imaginary time" (WW 271; cf. 277).

67. *The Views of Sommerfeld, Cassirer and Weyl*

Others, however, are less cautious, and a great many relativists express themselves in a way suggesting that a profound tendency toward Minkowski's interpretation is inherent in the doctrine. Sommerfeld, for example, whose considerable works are well-known, particularly the admirable theory by which he succeeded in connecting the theory of relativity with quantum

70

theory (cf. §127 below), explicitly subscribes to Minkowski's view: "There is nothing in what Minkowski says that must now be withdrawn."[3] Among the philosophers, Cassirer expresses an altogether analogous opinion, stating that the assumption formulated by Minkowski in the above quotation "seems ... now to be strictly realized" (ER 92; Eng. 424).[4] Weyl's attitude is more ambiguous. He insists on the distinction between the three properly spatial dimensions and the temporal dimension and would have us speak rather of a universe with (3 + 1) dimensions. However, upon other occasions he himself uses the expression he seems to have excluded and, even more significantly, declares that

however deep the chasm may be that separates the intuitive nature of space from that of time in our experience, nothing of this qualitative difference enters into the objective world which physics endeavors to crystallise out of direct experience. It is a four-dimensional continuum, which is neither 'time' nor 'space'. Only the consciousness that passes on in one portion of this world experiences the detached piece which comes to meet it and passes behind it, as *history*, that is, as a process that is going forward in time and takes place in space.

Weyl is perfectly aware to what extent this conception upsets all our notions of reality. Nevertheless, our immediate feelings to the contrary, he believes his doctrine to be sound, for it is "only now [that] the theory of relativity has succeeded in enabling our knowledge of physical nature to get a full grasp of the fact of motion, of change in the world." He is so convinced of this that he finds it "remarkable" that three-dimensional geometry seems so obvious, while four-dimensional geometry is so difficult.[5] Minkowski clearly did not overstate the case when he used the word 'radical' to describe the implications of the relativistic conception.[6]

68. *The Views of Einstein, Eddington and Cunningham*

Moreover, Weyl is far from alone in his position. Quite apart from those like Sommerfeld who support Minkowski's statements without reservation, we see that Einstein himself asserted that in relativity, "from a 'happening' in three-dimensional space, physics becomes, as it were, an 'existence' in the four-dimensional 'world'."[7] Eddington maintains that in relativity theory "the continuum formed of space and imaginary time is completely isotropic for all measurements; no direction can be picked out in it as fundamentally distinct from any other." Thus "events do not happen; they are just there, and we come across them [as we follow our world line]. 'The formality of taking place' ... of the event in question ... has no important significance" (STG 48, 51).[8] According to Henri Marais,

One can say that the separation into time and space is evidently an artificial process creating the 'illusion' of succession and expressing some sort of relationship – fundamental for us, accidental from the standpoint of reality – between our private perception of the world and the world line of our organism.[9]

Cunningham's statement also brings out this aspect of the theory in a very clear and straightforward way:

The distinction [between space and time] as separate modes of correlating and ordering phenomena is lost, and the motion of a point in time is represented as a stationary curve in four-dimensional space. ... The whole history of a physical system is laid out as a changeless whole.

He also writes:

There is perhaps an analogy to be drawn between the analysis which lays out the whole history of phenomena as a single whole, and the things in themselves, the natural phenomena apart from the human intelligence, for which consciousness of time and space does not exist ... [and] in which, as far as they are mechanically determinate, the past and the future are interchangeable. Such a view of the universe is ... the view of an intelligence which could comprehend at one glance the whole of time and space.

But the limitations of the human mind resolve this changeless whole into its temporal and spatial aspects, and the past and future of the physical world is the past and future of the intelligence perceiving it.[10]

69. *The Spatialization of Time in Relativism*

It should be noted that if time and space are henceforth to be more or less merged into a single continuum, this change will clearly work to the advantage of space. This is apparent from the evidence we have just cited, for if becoming is to be transformed into being (according to Einstein), so that the act of taking place becomes a simple unimportant formality for an event (according to Eddington); if succession is only an illusion (according to Marais), and if every physical system constitutes a changeless whole (according to Cunningham) – this can mean only one thing: the abolition or disappearance of time. Therefore, Cunningham does not hesitate to speak of "the timeless universe of Minkowski" (*Principle* 214). Let us observe, moreover, that this already follows from the very fact that the construction at which one arrives is a *geometry* (§62). And one need only open an exposition of the doctrine to note that, where time is concerned, the writer always speaks of one dimension, obviously conceived as spatial, while no attempt is ever made to represent the properly spatial dimensions in terms of time. Thus Marais, in the preface to the fine book we have already had occasion to quote in several contexts, does not hesitate to affirm that relativity aims at "incorporating time into space" (*Intro. géométrique* vi), and his testimony carries all the more weight because Marais, whose essay is entirely mathematical, is also, as we know, a very competent philosopher.

70. *The Irreversibility of Phenomena*

Sometimes this tendency to assimilate time with space – which is really, as we have just noted, a transformation of time into space – seems to be carried to extremes. Not only does it go farther than would seem to be authorized by our immediate sensation (a consideration for which the relativists rightly care very little), it even exceeds the authority of the most clearly established facts and the most basic foundations of science. Can we really dissolve time into space as Minkowski assumed? Is it accurate to say with Eddington that in physical reality "the continuum formed of space and imaginary time is completely isotropic for all measurements" and that "no direction can be picked out in it as fundamentally distinct from any other"? On the contrary, it is clear that, taken literally, these are completely extravagant propositions having no connection with phenomena. The temporal dimension is by nature different from the spatial dimensions; we know this with sure and immediate knowledge, with a certainty that cannot be shaken in the least by any rational arguments, no matter how seductive. In fact, in the case of the spatial dimensions, we can largely move about in them at will; this is the axiom of *free mobility*, to use Bertrand Russell's phrase,[11] and if we but ask ourselves, we shall realize that it is an integral part of our idea of space. In the same way we can ascertain that the notion of time contains no such element. Admittedly it would no longer be quite exact to state, as was done before Einstein and Minkowski, that we all progress continually along the temporal dimension in one and the same direction with a necessary and uniform movement. Langevin, in one of the admirable expositions he devoted to the theory, showed that a traveler who left the earth at half the speed of light would find upon his return that two centuries had passed, while he himself would have aged only two years. But it is nonetheless certain that the ability to move in time is extremely limited, even according to the new concepts themselves. Of course, relativistic physicists are perfectly aware of this and make the necessary qualifications if they are examining these questions closely. Thus Einstein himself uses the argument that we "cannot telephone into the past," and Eddington states that the notion of entropy has survived the Einsteinian revolution and is (along with the principle of least action) one of the two generalizations toward which physics is converging (STG 149). Now the notion of entropy, which grew out of Carnot's principle, is nothing more than the expression of the irreversibility of phenomena, that is, of continual progress in time. And this is certainly the immediate conviction our consciousness gives us as it creates the notion of time – *certa scientia et clamante conscientia*, according to the scholastic expression Maine de Biran was fond of quoting. This conviction is continually confirmed by countless observations: we *know* that the chicken will not go back into the egg and that we are not any more likely in Einstein's world then in Newton's to walk in reverse or to digest before we have eaten. After the passage we quoted above (§67) Wien goes on to say: "Neither the theory of relativity nor any other theoretical concept can alter the fact that time is something totally different from a spatial dimension."

71. *The Source of the Relativistic Exaggerations*

How does it happen that so many authoritative accounts of the theory of relativity would seem to imply the contrary? Sometimes they apparently claim that time is henceforth indistinguishable from space, since the four dimensions of 'Minkowski's world' are supposed to be perfectly isotropic. At other times, when they admit having reservations on this point (like Weyl, when he stipulates that one must not speak of the four dimensions of the universe but of 3 + 1 dimensions, thereby granting that the temporal dimension is not of the same nature as the spatial dimensions), these reservations are clearly insufficient, referring only to the fact that time seems to be qualified by an imaginary factor in Minkowski's formula, and not referring to the irreversibility we know to be so fundamental. Is this to be interpreted as an anomaly produced purely and simply by Einstein's theory, a sort of vicious tendency his doctrine inculcates in the minds of his followers? Not at all, since we need only examine a little more closely the evolution of our scientific knowledge to realize that, no the contrary, we are dealing here with a general tendency inherent in our reason, a tendency relativism is apt to render more visible by the very fact that it pushes its explanations so far.

72. *Identity in Time*

Common sense, although it recognizes that all things are subject to the conditions of time and space, does not treat time and space in quite the same way. By nature space appears totally indifferent to things: they can undergo no modification as a result of my having changed their location.[12] It is true that if I carried a puppy to the top of Mont Blanc he would suffer, and that if I plunged him into water he would be asphyxiated, but this is the result of a change in the visible material conditions of his environment and not the result of mere spatial change. On the other hand, by the very fact of advancing in time he will undergo modification: he will become older. If twenty years from now someone showed me a dog exactly like this one and tried to make me think it was the same one, I would certainly not believe him.

Our reason, however, does not remain fixed in such an attitude. On the contrary, it seeks to *explain* the modifications time brings to things, which means that, all things considered, our reason assumes that there should be no change simply as the result of the passing of time. What is behind this search for explanations or *causes* is, therefore, a conception that makes objects indifferent to their displacement in time, that is, treats temporal displacement as if it were displacement in space. It is just as obvious that as soon as we bring time into our calculations, even if it is only for the purpose of simple prediction, we must to some extent yield to the same tendency. For we represent time by a symbol, as a magnitude. What characterizes magnitudes is that they can increase or decrease — while time never goes backward, and only in our imagination can we endow it with regressive movement.

73. *The Spatialization of Time in the Past*

Nevertheless, modern physicists have seemed to take the assimilation of time into a spatial magnitude almost for granted. Thus for Descartes time is a *dimension*, and he defines the latter term as follows: "By dimension I understand nothing but the mode and aspect according to which a subject is considered to be measurable."[13] D'Alembert considers that "without knowing time in itself and without even having a precise measure of it, we cannot represent the relationship between its parts any more clearly than by the relationship between the segments of an indefinite line." Elsewhere he writes: "A clever acquaintance of mine believes that duration could be regarded as a fourth dimension."[14] Lagrange claims that "mechanics can be regarded as a four-dimensional geometry, and mechanical analysis as an extension of geometrical analysis."[15] Moreover, scientists have not been the only ones to follow this natural turn of mind. As Léon Brunschvicg rightly points out,

modern philosophy has generally presupposed a correspondence between space and time. Spinoza's *Letter to Louis Meyer*, the formulations in the *Leibniz–Clarke Cor-*

respondence, and the key notions of the *Transcendental Esthetic* show the extent to which time has shared the destiny of space in classical rationalism (EH 499).

Thus the relativists are only extending — rather far, to be sure — a trend characteristic of science: It is quite simply the perennial attempt to explain becoming, change in time, by the negation of this change. Here their theory is, as Bergson says,

the metaphysic immanent in the spatial representation of time. It is inevitable. Clear or confused, it was always the natural metaphysic of the mind speculating upon becoming.[16]

Indeed, more or less from the beginning of Greek philosophy, this thought finds an expression — in many respects the definitive expression — in Parmenides' image of the sphere. But it is frequently found, in the most diverse forms, to this very day. Hegel approaches it throughout his work and comes so close to it that one of his disciples and most faithful followers [J. M. E. McTaggart], rigorously developing the master's teachings, makes it almost explicit (cf. ES 2:63 ff.).

74. *The Dissymmetry between Time and Space*

Furthermore, in the absence of any other proof, the very exaggerations of the relativists and the fact that they can sustain the illusion of having accomplished what no theory could possibly accomplish — namely, the complete assimilation of time into space — would be sufficient to demonstrate this. It is also demonstrated by the related, but less controversial, fact that this assimilation, where the new theory achieves it (more or less incompletely, to be sure), seems to them to be significant progress. Langevin, for example, stresses the point that "Galileo's [equations], which characterize ordinary kinematics, introduce a dissymmetry between distance in space and the interval of time between two events, a dissymmetry that disappears in the new kinematics"; he seems to consider this a considerable advantage that the latter theory has to offer (PR 10).[17] Jean Becquerel goes even further, for upon observing that "space and time play different roles in the old conception" (which could not be more accurate), he adds: "We shall see this dissymmetry disappear in the Space—Time of the new theory."[18]

75. *Carnot's Principle and Plausibility*

Finally, it should again be noted here that the relativistic exaggerations in this area have usually been accepted without too much protest. Quite obviously this is the direct consequence of a state of mind much like the one that quickly accepted the long awaited laws of conservation as *plausible* [19] as soon as they were proclaimed, yet allowed everything concerning the irreversibility of phenomena to be ignored for so long that teaching of Sadi Carnot is still without an echo. Even in 1875, twenty years after the publication of Clausius's basic work, a thinker as penetrating and well-informed as Cournot was able to proclaim that the only thing involved in the transformation of heat into kinetic energy was simple "waste products," constituting a "disturbance whose causal influence will continually diminish as the apparatus is perfected." [20]

Thus, there is a real analogy between the way the theory of relativity treats time and the way it treats gravitation. In both cases, concepts that were in no way geometrical for either common sense or pre-Einsteinian physics are formulated in geometrical terms, that is, are reduced to geometry. An attempt is made to spatialize them: to grasp them, understand them and explain them by means of spatial concepts.

NOTES

[1] H. A. Lorentz, A. Einstein and H. Minkowski, *Das Relativitätsprinzip* (Leipzig, 1913), pp. 56, 59 [Hermann Minkowski, 'Space and Time,' *The Principle of Relativity*, trans. W. Perrett and G. B. Jeffery (London: Methuen, 1923; reprint New York: Dover, n.d.), pp. 75, 80].

[2] Max Born, *La théorie de la relativité d'Einstein*, trans. Finkelstein and Verdier (Paris, 1923), p. 283. For the importance of Minknowski's works, cf. also Max von Laue, *Die Relativitätstheorie* (Brunswick, 1919), 1:118, 169, 196; and Ebenezer Cunningham, *The Principle of Relativity* (Cambridge, 1914), p. 86.

[3] Lorentz *et al.*, p. 69 [Arnold Sommerfeld, Notes to Minkowski's 'Space and Time,' *The Principle of Relativity*, p. 92]. Cf. also what we have said on this subject in ES 2:377, n. 3.

[4] Cf. ER 119 (Eng. 449): "All spatial and temporal values [are] exchangeable with each other. . . . The direction into the past and that into the future are distinguished from each other . . . by nothing more than the + and − directions in space."

[5] STM 283, 217 (a part of the first quoted passage is cited in §47 above). It can be seen (STM 274, 283) that the author does not shrink from the very strange consequences of these conceptions.

[6] Lorentz et al., p. 56 [Eng., 'Space and Time,' p. 75].

[7] Cf. Roberto Marcolongo, *Relatività* (Messina, 1921), p. 98 [RSG, 122]. In a more recent work Einstein declares that "it is neither the point in space, nor the instant in time, at which something happens that has physical reality, but only the event itself," so that "there is no absolute . . . relation in space, and no absolute relation in time between two events, but there is an absolute . . . relation in space and time." He adds

that "the circumstance that there is no objective rational division of the four-dimensional continuum into a three-dimensional space and a one-dimensional time continuum indicates that the laws of nature will assume a form which is logically most satisfactory when expressed as laws in the four-dimensional space–time continuum." At the same time he does recognize, however, that we "must remember that the time co-ordinate is defined physically wholly differently from the space co-ordinates" (VVR 20–21; Eng. 30–31).

8 [The bracketed phrase occurs at this point in the translation used by Meyerson (*Espace, temps, gravitation*, trans. J. Rossignol, Paris, 1921, pp. 59, 63), but not in Eddington's original. Although it is certainly true that many interpreters of relativity did – and many still do – interpret the theory as these quotations from Eddington suggest, Eddington himself did not. If one reads STG carefully, the context makes it clear that the position Eddington is describing here is one he himself rejects. Cf. Eddington, *The Nature of The Physical World* (Ann Arbor: University of Michigan Press, 1958), pp. 50–52, 55–58; Milič Čapek, *The Philosophical Impact of Contemporary Physics* (New York: van Nostrand, 1969), p. 186, n. 9.]

9 Henri Marais, *Introduction géométrique à l'étude de la relativité* (Paris, 1923), p. 96. This statement, significantly enough, follows those cited above §§20 and 47, where he affirms the reality of the entities defined by physical theory in general and relativity theory in particular.

10 Ebenezer Cunningham, *The Principle of Relativity*, pp. 191, 213–214 [the first bracketed insertion is Meyerson's].

11 Bertrand Russell, *An Essay on the Foundations of Geometry* (Cambridge, 1897), p. 150, §144 [Meyerson quotes *Essai sur les fondements de la géométrie*, trans. Cadenat (Paris, 1901)].

12 Here, and in the following paragraphs, we are only summarizing our arguments in IR 29 ff. (Eng. 37 ff.), ES 1:150 ff., and at the 6 April 1922 meeting of the *Société Française de Philosophie* (*Bulletin* 22 [1922] 107 ff.) [See Appendix 2].

13 Descartes, *Rules for the Direction of the Understanding*, rule 14 [*The Philosophical Works of Descartes*, trans. Elizabeth Haldane and G. R. T. Ross (Cambridge, 1931; reprint New York: Dover, 1955), 1:61].

14 Jean Le Rond d'Alembert, *Traité de dynamique* (Paris, 1758), pp. vii–viii, and *Encyclopédie* (Paris, 1751), under the word 'Dimension', 4:1010. The following is a more complete text of the second of these passages, which I find particularly interesting: "I have said above that it was not possible to imagine more than three dimensions. A clever acquaintance of mine believes, however, that duration could be regarded as a fourth dimension and that the product of time and solidity would be in some way a product of four dimensions; that idea can be contested, but it seems to me that it has some merit, if only that of novelty."

15 Joseph Louis de Lagrange, *Théorie des fonctions analytiques*, Œuvres (Paris, 1867–1892), 9:337.

16 Henri Bergson, *Durée et simultanéité* (Paris, 1922), p. 82 [*Duration and Simultaneity*, trans. Leon Jacobson (Indianapolis: Bobbs-Merrill, 1965), p. 61].

17 [The bracketed insertion is Meyerson's.] Cf. PR 35.

18 Jean Becquerel, *Le principe de la relativité et la Théorie de la gravitation* (Paris, 1922), pp. 8–9; cf. p. 36. Edmond Bauer likewise insists on the fact that "in classical theory" there remains "a complete dissymmetry" between time and space, which

"somewhat compromises the rigor and elegance of classical kinematics" (*La théorie de la relativité*, Paris, 1922, pp. 23–24).

[19] See Ch. 4, n. 6.

[20] Antoine Cournot, *Matérialisme, vitalisme, rationalisme* (Paris, 1975), p. 93.

CHAPTER 8

ELECTRICAL PHENOMENA

76. *The Experimental Bases of Relativism*

In the final form Einstein gave it, relativity theory, as we have said, embraces all the phenomena attributed to gravitational action. On the other hand, it completely leaves aside everything having to do with the action of electricity. But perhaps further clarification is needed to prevent any possible misunderstanding.

It has been said that relativity grew out of the Michelson–Morley experiment, and this is historically accurate, for it is this experiment that made it absolutely clear that no perceptible consequences of absolute motion can ever be produced. Thus Wien is not mistaken in suggesting that Michelson's experiment was as important for relativity theory as the experiment demonstrating the impossibility of perpetual motion was for the law of the conservation of energy (WW 266). But it is clearly a mistake to conclude from this that the new theory has only a very narrow experimental basis. As Langevin points out, "Michelson's experiment is not an isolated one . . .; it is only an extremely precise verification of a theory, the electromagnetic theory," since the "phenomenon of the isotropic propagation of light conforms to the theories of Maxwell and Lorentz as established by the entire series of electromagnetic experiments." Actually, "the time implicitly introduced by the founders of electromagnetic theory, and likewise by all physicists working in this field each time they apply the fundamental Maxwell–Hertz laws in their ordinary form, is none other than relativistic time." This became apparent only after the fact, that is, after Michelson's experiment made it clear that classical mechanics had to be modified. But, as a matter of fact,

the whole detour taken by the Michelson experiment could have been avoided if one had really believed that the equations of electromagnetic theory represent all electromagnetic experiments and that the property these equations have of conserving their form under certain transformations represents the experimental fact of relativity (AG 11–12; PR 20).[1]

These transformations are the ones indicated by H. A. Lorentz, and they are opposed to those of classical mechanics. Thus, as Jean Becquerel explains, from a more general point of view Michelson's results appear to be "the result of a conflict between the laws of Newtonian mechanics and the laws of electromagnetism."[2] One must choose between them, and the Einsteinians have made their choice, for reasons that have been quite clearly explained many times, the principal one being that the laws of electromagnetism have been verified with great precision, while the laws of mechanics permit only a *grosso modo* verification. Consequently it is possible to consider the laws of classical mechanics valid as long as we are dealing with phenomena of sufficient magnitude (as relativity theory, moreover, allows), while at the same time assuming that they become inapplicable and must give way to the laws deduced from the observation of electrical phenomena as soon as more precise observations are involved.

77. Relativism and Optics

Nor, we might add, is it meaningful to weigh (as has sometimes been done) the small number and minute detail of the experimental results that gave rise to relativism against the huge number of observations accounted for by optics. Fresnel's theory, it is true, does continue to be expounded in nu-merous and voluminous text books, where it admittedly serves a very useful purpose. But from the point of view of actual theoretical physics today, this hypothesis has, strictly speaking, ceased to exist as an independent concept. The equivalence of luminous and electrical waves suspected by Maxwell has been demonstrated by Hertz and Rubens, among others, causing optics to become, with Lorentz's work, a subdiscipline of the physics of electricity.[3] Now this physics, far from being opposed to the relativistic hypothesis, actually serves as a basis for it, as we have just seen. Thus, what is really at stake for those who support Fresnel against Einstein is not the relativistic hypothesis but the equivalence of optics and electricity (already firmly established in the last years of the nineteenth century).

78. Einstein's Theory does not Explain Electricity

Since relativity theory is, therefore, the confirmation — the very embodi-ment — of the triumph of the conceptions arising out of the study of electro-magnetic phenomena, it goes without saying that there could not be the least contradiction between them. Thus Einsteinian relativity *applies* without exception to everything pertaining to the action of electricity; however, this does not mean that it *explains* these phenomena in the sense that it explains gravitation. For in the case of gravitation, as we have just seen, once one has established the presence of a body — that is, the existence of a particular structure in that part of space — all gravitational phenomena logically follow from it. As soon as electrical phenomena intervene, on the other hand, we must assume a particular activity in Einsteinian space that does not follow from the nature of that space alone. Let us recall here the distinction we drew above (§43) between theory as abstract principle and explanatory theory, the former intended to sum up a collection of lawlike relationships, and therefore itself sharing in this lawlike quality, and the latter setting out to make the phenomenon comprehensible to the intellect, to make it rational. Obviously the same distinction applies here. Einsteinian relativity exhibits both of these characteristics at the same time; consequently it is able to deal with electrical phenomena in its capacity as a simple abstract principle, while in the case of gravitational phenomena it plays the role of a genuine explanatory theory.

79. The Prime Phenomenon

However, the great success of Einstein's concepts necessarily encouraged theorists to continue along the path he had opened up and to include electricity in the system of relativistic explanations. This was all the more compelling because, assuming that the result sought by Einstein had been achieved and all phenomena having to do with gravitation had been deduced, the only mystery remaining in physics was that of electricity (except for problems arising out of the discontinuity of matter, which we shall treat below). Moreover, this was an absolutely fundamental problem, since, as we are well aware, modern physicists consider the electrical phenomenon to be the archetypical phenomenon, the prime phenomenon, what a *Naturphilosoph* would have called the *Urphänomen*, that is, the phenomenon to which all others must finally be reduced. It must be admitted, however, that on closer examination this situation seems less clear-cut than it would appear on the surface.

We have discussed (§41) the way in which physicists' concepts of the nature of this prime phenomenon have changed. But we believe it will be useful to recall how recently these changes occurred. In spite of Hume's demonstration and the resulting impossibility of formulating any kind of coherent and logical theory of impact, science nevertheless persisted for the greater part of the last century in considering this phenomenon rational, that is to say, not only explicable but completely explained. This is a position

Cuvier unqualifiedly affirmed hardly a century ago.[4] Physicists might be inclined to doubt the weight of this testimony, given that it comes from a biologist, but there is no doubt that this great man was completely abreast of all the scientific achievements of his epoch. In this case he was only expressing the common belief of contemporary scientists, as is demonstrated by the fact that more than a generation later, at the beginning of the second half of the century, Maxwell (although he had, following Faraday, laid down the basic principles underlying the modern conception of electrical phenomena) still devoted all the resources of his genius, if not actually to reduce electricity to mechanics, at least to establish in each particular case that such a reduction was possible. His principal work, the *Treatise on Electricity and Magnetism*, is entirely imbued with this spirit, as Henri Poincaré observes.[5]

No change in attitude really occurs until the last years of the nineteenth, or even the beginning of the twentieth, century, scarcely a generation ago. And it is quite interesting to observe that there is already something like a step backward, or a reaction against the electrical theory, a reaction due to Einstein's theory. This statement may seem strange, given what we know about the origins of relativity, growing as it did out of the equations summing up the laws of electricity. But this can be explained by the fact that here again we are concerned with the purely lawlike aspect of the theory. In its role as an explanatory theory, on the other hand, it conflicts with, or at least competes with, the electrical theory, in the sense that it makes the latter superfluous whenever the simple structure of space is sufficient to account for the observed phenomenon. "The case for the electrical theory of matter," Eddington aptly says, "is actually weakened, because many experimental effects formerly thought to depend on the peculiar properties of electrical forces are now found to be perfectly general consequences of the relativity of observational knowledge" (STG 61). One might say that the effects to which Eddington refers, which had been transferred from the domain of mechanics to that of electricity, were once again incorporated into mechanics by Einstein's theory.

80. *Explanation of Electrical Phenomena*

However opinions may change as to which of the two is more fundamental, modern physicists certainly believe that mechanics and electricity between them account for the whole realm of physical phenomena. This helps us understand how tempting it must have been for supporters of Einstein's theory to extend the method he had used so effectively, that is, to seek to explain electrical phenomena by means of a particular structure of space, or in other words, to transform the physics of electricity into a pure geometry.

The best-known attempt to accomplish this was that of Hermann Weyl. His work was completed by Eddington, whose name we have so often had occasion to cite; other physicists also participated in this effort. What generally characterizes these attempts is the fact that they end up attributing to space a structure even more complex than the one suggested by Einstein. Instead of the ten parameters attributed to space by Einstein's theory, Weyl's conception needs fourteen. It goes without saying that (even less here than in the case of the theory of relativity itself) we do not take it upon ourself to explain to the reader the mystery behind these terms, and, just as in the case of Einstein's theory, we claim no competence to judge whether or not the conception extended in this way is acceptable from the scientific point of view. From such a standpoint, the situation seems much less clear-cut than in the case of the theory of gravitation. Although he finds Weyl's effort to be "an exceedingly ingenious attempt," Einstein does not believe that the view of his continuator "will hold its ground in relation to reality" (ER 23). Furthermore, Weyl himself admits that this theory is still in its infancy,[6] and Langevin seems to see it as only a beginning, since he limits himself to expressing the hope "that into this [relativistic] geometry we shall be able to incorporate electromagnetism, which now constitutes physics" (AG 22).[7]

81. *Purely Geometrical Reality*

Here, for the sake of argument, we shall consider the problem solved. Suffice it to say that anyone who has grasped the meaning of these deductions and has been able to stretch his imagination far enough to comprehend the reality of this extremely complex structure of space will understand not only the how of physical phenomena (laws alone could accomplish this) but also the why, that is, the cause of all the properties (with a few exceptions we shall treat later) that characterize the real world as it is made known to us both in immediate perception and in perception refined by scientific instruments. This is because, as Eddington rightly goes on to point out after the passage just cited above, with "what remains of physics ... all the other sciences" will be included in the relativistic deduction, constituting a more general geometry than Einstein's.[8]

The reader will now be better able to appreciate the appropriateness of Borel's statement characterizing the relativistic explanation, unlike those that preceded it, as *geometrical*. In relativism, taken to its logical conclusion, *everything* is geometry and only geometry, while in mechanical theories geometry is simply applied to concepts of a nongeometrical nature, such as the chemical atom, the material particle, etc.

NOTES

[1] Einstein himself, moreover, has explicitly stated that the theory of relativity has "been developed from electrodynamics as an astoundingly simple combination and generalisation of the hypotheses, formerly independent of each other, on which electrodynamics was built" (RSG 41; cf. 44, 49).

[2] Jean Becquerel, *Le principe de la relativité et la Théorie de la gravitation* (Paris, 1922), p. 35. Emile Borel (ET 170; Eng. 143) states that "the most complete proof ... of the theory [of relativity] results from the study of the laws of electrodynamics as a whole, and from the fact that the system of equations governing these laws is invariant for the group of Lorentz transformations" [the bracketed insertion is Meyerson's]. Weyl points out that "in treating electrodynamics after the manner of Maxwell, one was already unconsciously treading in the steps of the principle of relativity" (STM 205). Moreover, this physicist goes further, writing: "We have ... a good right to claim that the whole fund of experience which is crystallised in Maxwell's Theory weighs in favor of the world-metrical nature of electricity" (STM 284), that is to say, in favor of the concepts developed by Weyl himself.

[3] Lucien Fabre's *Une nouvelle figure du monde: Les théories d'Einstein* (Paris, 1921, p. 93 ff.) contains a lucid account in nonmathematical language of the relationships between the theories of Fresnel, Faraday, Maxwell, Hertz, and Lorentz.

[4] Georges Cuvier, 'Histoire des progrès des sciences naturelles' in *Œuvres complètes de Buffon, Complément* (Paris, 1862), 1:2.

[5] Henri Poincaré, *Electricité et optique* (Paris, 1901), p. 3 ff.

[6] Weyl, *Raum, Zeit, Materie*, preface to 3rd ed., reproduced in the 4th, but omitted in the [French] translation [STM x–xi].

[7] [The bracketed insertion is Meyerson's.] Cf. also Weyl's opinion in §105 below.

[8] [We cannot locate this quotation in either the English original or the translation Meyerson uses (*Espace, temps, gravitation*, trans. J. Rossignol, Paris, 1921).] Weyl himself, moreover, observes that, according to his theory, "the electromagnetic field and the electromagnetic forces are then derived from the metrical structure of the world," adding: "No other truly essential actions of forces are, however, known to us besides those of gravitation and electromagnetic actions; for all the others statistical physics presents some reasonable argument which traces them back to the above two by the method of mean values" (STM 283).

CHAPTER 9

BIOLOGICAL PHENOMENA

82. *Mental Phenomena*

It is quite likely that Eddington had in mind only the physicochemical sciences when he expressed the opinion cited at the end of the last chapter.

Is it conceivable, however, that the theory of relativity will eventually encompass phenomena outside this area? This possibility certainly seems difficult to accept in the case of all those branches of knowledge the Germans call *sciences of the mind* (*Geisteswissenschaften*).

It is true that opinions have been expressed to the contrary. Petzoldt, for example, states that it is "absolutely inevitable that physics will be reconciled with the physiology of sensations through relativity theory," and that in this way "the unjustified opposition between the physical and natural sciences on the one hand and the sciences of the mind on the other will be set aside" (SR 121).[1] But this can be explained by the fact that, as we have already noted, this disciple of Mach, in his attempt to reduce relativism to a completely orthodox positivism and phenomenalism, is convinced that Einstein's theory connects the phenomenon more closely to the self than does common sense or pre-Einsteinian physics. If, on the contrary, one keeps in mind that, as we think we have established, the theory of relativity is continuing the process of eliminating anthropomorphism described by Planck (§§22 and 48), and if one considers the representation of reality at which the new theory arrives, it would seem impossible not to see that just the opposite has occurred: the gulf between these two branches of human knowledge has become even wider.

83. *Vital Phenomena*

At first glance it seems nearly as hard to believe that relativity theory could be used to explain biological phenomena. Can one hope to transform the striking diversity life presents at every moment into a purely spatial diversity? But our very formulation of the question shows that we are dealing here with a completely general problem. The only reason it so imperiously commands our attention here is that through relativity theory the reduction of physics

86

to space has largely been accomplished. For the real problem is nothing other than the reduction of the vital to the physical, a reduction the antivitalist aspires to achieve and the vitalist claims to be impossible. We have addressed this essential question in Chapter 7 of *De l'explication* [ES], where we tried to distinguish which claims of the two opposing camps can be justified. We shall only point out here that, insofar as manifestations of life can be assimilated with those of non-living matter, they will clearly come under the jurisdiction of the relativistic conceptions, that is, however paradoxical this may seem to us, they will be reduced to purely spatial concepts.

84. *The Vital and Hyperspace*

One might go even further in this direction and suppose that higher geometry will perhaps some day be able to be used for biological explanations more or less directly, that is, without going through physics. At first this idea seems strange. However, leaving aside the use (or abuse) that has been made of the concept of hyperspace in seeking to make it the basis for theories of certain psychic or *metapsychic* phenomena, it should be recalled that, more than a generation before the appearance of Einstein's theory, Ernest Renan asserted that "what modern geometry has said about space having more then three dimensions may have some connection with reality," given that one could thus understand how "the types of generation are contained within one another."[2] The fact that so perceptive and judicious a thinker could take this sort of possibility seriously would seem to be sufficient cause in itself for us not to dismiss the question without further debate.

NOTES

[1] Cf. Ch. 1, n. 6; Ch. 3, n. 5; and Ch. 5, n. 1, above, on this philosopher's attitude.
[2] Cf. Georges Sorel, 'Vues sur les problèmes de la philosophie,' *Revue de métaphysique et de morale* 18 (1910) 610.

CHAPTER 10

UNIVERSAL EXPLANATION

85. *Relativism as a System of Universal Deduction*

For the time being let us disregard these very remote and speculative views and consider the structure of the theory as it is viewed by most of its adherents. If we attempt to survey the whole theory all at once, we cannot fail to be struck by the breadth of ambition that seems to have consciously or unconsciously motivated those who constructed it. For what they sought to establish was nothing less than a true system of universal deduction, in the sense that Cartesian physics or the natural philosophy of Hegel constitutes such a system.

86. *Relativistic Geometry is Still Deductive*

Relativists certainly speak at times as if what they were doing was simply generalizing from experimental data: they call their geometry *natural* and claim to be content to observe the way bodies behave in space. Moreover, this is not an altogether empty claim. It is accurate, in fact, to say (as we did in our preface) that in the end the fate of Einstein's theory and that of the concept of space and time it implies depend on physical data. But does this justify the claim that the new geometry is an essentially different discipline from classical geometry in terms of its relationship to experience? It is clear, on the contrary, that there has been no essential change and that modern geometry remains just what classical geometry was: a body of knowledge in which deduction plays the predominant role. As we noted in the case of Euclidean geometry (§40), experience only comes into play in the case of the initial axioms and postulates; in this particular case, it merely suggested which system to adopt. But the system itself is nonetheless a geometry, that is, deductive to the extent that any geometry must be deductive.[1] Thus, insofar as this geometry will come to replace physics, as is suggested by the evidence we have cited, it will aim at a complete deduction of the phenomena in its proper domain, exactly as does the system of Descartes or Hegel.

88

87. *Relativism and Descartes's System*

In the case of Descartes, the resemblance strikes us immediately. Indeed, for Descartes too, all physical reality is only geometry; geometry is at the center of everything and is sufficient for everything. Furthermore, Weyl himself notes this fact: "Descartes's dream of a purely geometrical physics seems to be attaining fulfilment in a manner of which he could certainly have had no presentiment."[2] If there is any overstatement here, it is in the distinction Weyl seems to make between the Cartesian theory and the current theory. For the former was not just a dream, and it was much more than a program — the program all science eternally sets itself — it constituted an actual enterprise undertaken with the aid of all the resources the science of that time was able to make available to a genius uniquely gifted in both science and philosophy. This enterprise finally ended in failure — at least it would seem so from our present standpoint — but it is certain that for nearly two generations it appeared to the majority of competent men of science to be a great success.[3] Now, we do not know what posterity will think of the theory of relativity a half century from now, but it would not be exceeding the bounds of fair judgment to say that what it has accomplished is perhaps less complete and certainly less definitive than Weyl's words would seem to suggest.

88. *Relativism and Hegel's System*

The analogy between relativism and the Hegelian philosophy of nature seems at first much less obvious, and we can produce no evidence for our claim from either a supporter or an adversary of the doctrine. However, we believe that here too the resemblance is very real and dare hope that the reader will be convinced by the pages that follow. Indeed, we shall return to this point more than once, for the comparison between the deductions of the relativists and those of Hegel seem to us to serve especially well to reveal the implications of the modern deduction. This is precisely because the attempt of the German philosopher, although it is closer to us in time, proceeds in a manner much farther from our habitual way of thinking than that of Descartes (§25). For this reason, and also because his procedure seems much less successful than Descartes's, his devices are less apt to take us in and thus become more apparent.

89. *The Limits of the Three Systems*

If we begin by considering the limits of these three systems of universal deduction of the physical world, we shall see that the relativistic deduction is more comprehensive than Descartes's, for it goes beyond the bounds the latter had set for his geometrical explanations. As a matter of fact, Descartes did not intend to include anything dealing with time, accepting the notion of time, like the notion of space, as presented to him by common sense perception. In this respect the new universal deduction is actually more like that of Hegel, who intended to go so far as to deduce the three-dimensionality of space. One must admit, however, that when it comes to the question of how large an area each of the two systems would comprehend, the dissimilarity between relativism and Hegelianism at first appears to be so great that it would seem to exclude all analogy. This is a result of the fact that over the years a sort of distortion has occurred in the way the Hegelian system is perceived. In fact, Hegel was not content to recreate physical reality; he also intended to 'construct' the totality of man's spiritual activity in the same way. Death interrupted him in the midst of this truly gigantic project, but he nevertheless had the time to complete the philosophy of right, the philosophies of history and religion, esthetics, and the history of philosophy. Now, strangely enough, albeit understandably, these superstructures of the doctrine have caught the imagination much more than what was meant to constitute the foundation for the entire edifice. Hegel's disciples were soon neglecting the philosophy of nature (today some, who can otherwise be counted among the most faithful, even explictly abandon it). Sometimes they even skipped over the master's logic in order to devote themselves to showing how *idea* tended to realize itself in this or that particular realm of intellectual evolution; it is certainly in this area that Hegel's influence has been most strongly felt. If, however, one looks not at the disciples but at the master himself, one is forced to reconsider. For Hegel, when giving a 'basic outline' (*Grundriss*) of his work as a whole in the *Encyclopedia*, first treated the *Logic* in one volume, then devoted a second volume to the *Philosophy of Nature* and a third to the whole of the *Philosophy of Spirit*, thereby indicating that he considered this third work to be, at most, equal in importance to the one that preceded it. In fact, the *Philosophy of Nature* is even more voluminous than the *Philosophy of Spirit* (and this seemingly trivial detail is not without importance for a man as systematic as Hegel). Therefore, the Hegelian deduction really is primarily concerned with physical reality and is thus comparable to the deduction of the relativists in this respect.

The relativistic deduction remains well within the limits of Hegel's philosophy of nature in that it does not deal, as the latter does, with sensation: neither Einstein nor his continuators attempt to explain that which exists *sui generis* in our impression of the color red. For them, the light that produces this impression is something anonymous, something able to produce in us sensations totally unrelated to those of any color at all. The relativists are, above all, physicists, and as such they subscribe to the Democritean postulate precluding anything having to do with the quality of sensation. But in another sense, precisely because it is the work of scientists, the relativistic deduction goes well beyond the limits of Hegel's deduction. For Hegel intends to deduce only the most general characteristics of reality; the details, and in particular everything having to do with quantity, appear unimportant to him. He admits that "the idea of nature, in particularizing itself, breaks itself up into fortuitous things" which "depend on chance and spontaneity and are not determined by reason."[4] He thus marks out a marginal area containing a collection of phenomena with which deduction must not concern itself, and this area, as we have established, is extremely vast. It is so vast that it can, with a little imagination, include almost the whole domain Auguste Comte defined as that of positive science, as it was to be constituted according to his scheme (cf. ES 2: 128). The relativistic theory, on the contrary, means to deduce phenomena down to the last detail, and certainly their quantitative relationships above all. It grew out of observations concerning phenomena that are extremely subtle and difficult to detect, and it is in explaining these observations that it achieves its greatest triumph.

90. *The Universality of the Relativistic Deduction*

The relativists themselves do not hesitate to proclaim the universal nature of their deduction. Eddington, for example, states that a "geometer like Riemann might almost have foreseen the more important features of the actual world," these important features including, as his context makes clear, everything having to do with inertia and gravitation, that is to say, the whole of purely mechanical phenomena. If we could entertain the least doubt as to what this statement implies, it would be dispelled by the rest of the passage, for the author continues: "But nature has in reserve one great surprise — electricity" (STG 167). In other words, if electricity did not exist — that is, if there were only mechanical phenomena in nature, or rather if electricity could be reduced to mechanics — *all* the important phenomena of nature, without any exception, could have been foreseen by a geometer of genius; they would all be deducible. They become deducible if we combine Weyl's extension with Einstein's original theory. With such an extended theory we can then, in Eddington's words, "discern something of the reason why the world must of necessity be as we have described it" (STG 196). Furthermore, Weyl himself, in the last pages of his book, taking "a complete survey" of the relativistic deductions, states that *"it no longer seems daring to believe that we could so completely grasp* the nature of the physical world, of matter, and of natural forces, that logical necessity would extract from this insight the unique [*eindeutig*] laws that underlie the occurrence of physical events."[5] It can be seen that we did not exaggerate when we spoke of universal deduction in the case of relativity and compared this bold construct with that of Hegel.

91. *The Return to Reality*

The similarity between the fundamental attitudes of the two schools sometimes gives rise to striking analogies in the way they think and even in the way they express themselves. Weyl, in the passage just referred to, declares that the relativist, having arrived at this perfect understanding, is "overwhelmed [*überwältigt*] by a feeling of freedom won"; he feels that "the mind has cast off the fetters which have held it captive." But then *"our road has scaled such a height that it becomes difficult to come back down to all the familiar phenomena with which reality surrounds us."*[6] Now let us examine the sentiments of a disciple to whom the Hegelian absolute has just been revealed after an arduous ascent (we shall see shortly why it would be pointless to seek analogous statements by the master himself):

When one has worked one's way through the *Logic* and the *Encyclopedia* for the first time, one feels as if one has reached the summit of a high mountain. The air is rarefied and it is hard to see. One is no longer in the world of men; all that one sees at first is abstraction piled upon formidable abstraction, a metaphysical wilderness where it would not seem that any living spirit could dwell. One wanders though Being and Nothingness, Becoming, Limit and Essence, unable to catch one's breath, not knowing if one will ever get back to level ground, to earth. Gradually one sees through the clouds; patches of light begin to break through, the fog dissipates. Infinite perspectives unfold before one's eyes; entire continents stretch out, taken in at a glance. One would think one had arrived at the summit of human knowledge and the ultimate perspective on the world.[7]

Leaving aside Taine's style and imagery, does not this beautiful passage seem an amplification, or rather a simple paraphrase (in anticipation, to be sure) of Weyl's statements?

92. *The Mind Rediscovered in Nature*

Of course the method followed by the new doctrine is quite different from the one chosen by the author of the *Phenomenology*, who, as we know, detested mathematics in general and mathematical deduction in particular. Nonetheless, they are both trying to do essentially the same thing: it is still reason seeking to recreate the world, to extract the world from the depths of the mind itself. Eddington is not insensitive to this: "Those laws of nature which have been woven into a unified scheme — mechanics, gravitation, electrodynamics and optics — have their origin, not in any special mechanism of nature, but in the workings of the mind" — which would indeed explain why we are finally able to extract them from it. And in a few eloquent sentences at the end of his systematic work, he exclaims (alluding to Defoe's famous tale):

We have found a strange foot-print on the shores of the unknown. We have devised profound theories, one after another, to account for its origin. At last, we have succeeded in reconstructing the creature that made the foot-print. And Lo! it is our own (STG 198, 201).

Those who are at all familiar with the writings of Schelling and the *Philosophers of nature* of his school will easily recognize, in this way of rediscovering the mind in nature, the very inspiration that motivated these precursors of the bold Hegelian venture.

In an earlier work (ES 2:147 ff.) we spoke of the sometimes exaggerated criticisms of Hegel made by Trendelenburg, a philosopher of the following generation. We found him in error when he reproached Hegel for wishing to demonstrate the *a priori* nature of knowledge arrived at *a posteriori*, and pointed out that this reproach, if it could have been justified, would have applied equally well to Descartes and Kant. But one could not hope to find a more compelling refutation of Trendelenburg's alleged demonstration than that furnished by the aspect of relativism that concerns us here. For the relativist, who had not at all set out to seek the *a priori* in nature, discovered it nevertheless, one might say, in spite of himself.

NOTES

[1] Cassirer agrees on this point (ER 103; Eng. 434).

[2] STM 284. Cassirer similarly observes that, by abolishing the dualism between space and matter, physics abandons Newtonian ways and returns to those of Descartes (ER 61; Eng. 396).

[3] Léon Brunschvicg is only expressing the general feeling of everyone who has taken the trouble to study the *Principles of Philosophy* when he says that it seems that "man could never have hoped to grasp the supreme law of the universe by capturing it in such a

simple formula, making so obvious the connection between the mind and things" (EH 229).

[4] This is how Eduard Gans summed up the thought of the master in his preface to the *Philosophy of History* (Hegel, *Werke*, Berlin, 1842, 9: vii).

[5] *Raum, Zeit, Materie*, 3rd ed., p. 262. In the 4th ed. this passage, p. 274 of the [French] translation [STM 311] has been considerably weakened. Since we have undertaken to scrutinize the way scientists think, we believe it is interesting to consider the primitive form of their thought [italics have been added by the translators to identify the material not included in the 4th German edition or in Brose's translation of the 4th edition, which reads instead: "If Mie's view were correct, we could recognize the field as objective reality, and physics would no longer be far from the goal of giving so complete a grasp . . ."; the German insertion is Meyerson's].

[6] *Raum, Zeit, Materie*, 3rd ed., p. 262 [STM 312; italics have been added to indicate material not in the 4th ed.; the German insertion is Meyerson's]. Cassirer, examining the results of the relativistic conceptions from a philosophical point of view, likewise states that when one arrives at the ultimate abstractions of the theory, "finally there seems no return to . . . the immediate data, to which the naive view of the world clings" (ER 114; Eng. 444), while Borel simply notes that "the claim that we can include in a few coefficients the infinite variety of the perceptible world with its boundless diversity of qualities may appear extraordinary" (ET 206; Eng. 173).

[7] Taine, *Les philosophes classiques du XIXe siècle en France*, 11th ed. (Paris, 1912), p. 132. Cf. below, §139, for the conclusion of this passage.

CHAPTER 11

MATTER

93. *Matter Resorbed into Space*

Where the paths of relativism seem to diverge the most from those followed by Hegel they come closest to those of Descartes. In both relativism and Cartesianism matter is resorbed into space. For Descartes this identification is patent and avowed; it constitutes the postulate on which the entire system rests, since it is explicitly stated (in the passage quoted in §6 above) that the nature of a body

does not consist in its being hard, or heavy, or coloured, or one that affects our senses in some other way, but solely in the fact that it is a substance extended in length, breadth and depth.

In other words, since it has only purely spatial qualities, matter can only be spatial in nature. The relativists are a great deal less bold and explicit about it, but the final result is certainly the same. Matter disappears. Indeed, relativity explains the action of matter by attributing it to a particular structure of space. Now, all we know about matter is its effects; we can only conceive it in terms of these effects — this is an obvious truism. As Schopenhauer said, the reality of matter is its effects.[1] Take them away and nothing remains. For the relativist, matter is at most what the philosopher calls an occasional cause. "Matter is a symptom and not a cause," says Eddington; this "seems so natural that it is surprising that it should be obscured in the usual presentation of the theory." Indeed, "we cannot locate the pucker [representing the gravitational influence of a mass] at a point; it is 'somewhere round' the point." For "a gravitational field of force is not an absolute thing, and can be imitated or annulled at any point by an acceleration of the observer or a change of his mesh system." All that can be granted in this respect is that "the presence of a heavy particle does modify the world around it in an absolute way which cannot be imitated artificially." But this modification itself is purely spatial. Space and matter are so completely merged that "it does not seem possible to draw any distinction between the warping of physical space and the warping of physical objects which define space" (STG 191, 85, 126).[2] Furthermore, it is clear that this situation is implied by the

95

very fact that physics is reduced to a geometry in the new theory, a fact of which the relativists are fully aware, as we have seen above in the quotations from Langevin, Eddington, and Weyl.

94. *The Relativist's Reservations*

Thus, it is all the more strange that Eddington should register surprise at this direct consequence of the theory. One might say that he had reservations of some sort. At least he makes every effort to protect the relativists against the charge that they sought a result he clearly sees as paradoxical. "We did not consciously set out to construct a geometrical theory of the world; we were seeking physical reality by approved methods, and this is what has happened" (STG 183). The relativist's reservations, or at least his surprise, can be explained quite easily. He worked as a physicist, simply seeking to account for experimental observations, that is, to modify his conceptions of reality so as to reconcile them with these observations. He thought it necessary to make this effort because, unless he was able to construct a concept of *totality*, to use Höffding's apt expression, it seemed to him his physics would lose all meaning. But in the course of this work he never ceased to be profoundly conscious of the reality of the things with which he was concerned; at no time, he is convinced, did he yield to the temptation to explain physics by means of scientifically unreal concepts. Such a mode of explanation would be appropriate, as Hartmann puts it, only for the inhabitants of an asylum.[3] On the contrary, he was always careful to replace one reality with another reality, or at least something that seemed real to him. Nevertheless, at the end of his reasoning he finds himself faced with something that, no matter what he does, cannot be made to seem real to the same degree. For it is nothing but space, and however complicated he makes this notion, he finds it difficult to *thicken* it, so to speak, so that it becomes equivalent to the concept of matter. The relativist somehow has the impression of having performed a very difficult feat of prestidigitation: the pea has mysteriously disappeared. Only in this case it remains a mystery for him as well, and he swears that he never meant to make it disappear; it vanished all by itself.

95. *The 'Approved Methods' of Physics*

Actually, there is no mystery here at all, and nothing could be more natural than what has happened. The relativist is quite right to declare that he has done nothing but use the 'approved methods' of physics; but these very methods are what finally necessitate the disappearance of reality, its reabsorption into space. The physicist is not aware of this because his deductions are generally only quite limited in scope, and he explains reality only step by step. But he always explains it in spatial terms; he is constantly dominated by the concern to reduce all diversity to a purely spatial diversity. Consequently, if he should happen to sum up his efforts and combine his results in a more ambitious concept, he still necessarily finds only that which is spatial. His deductions were the small change of rationality; in adding them together he has merely consolidated his assets.

96. *Relativistic Space and the Hegelian Categories*

For this reason, in spite of formidable and seemingly quite irreconcilable differences, the way the relativist manipulates and in some sense molds space to produce something that, in the very process of accounting for the effects of matter, will render matter superfluous or useless and thus make it disappear, nevertheless strongly resembles the process by which Hegel seeks to transform his mental categories in such a way that they "take flesh and blood and walk into the air," in Seth's ingenious words (cf. §25 above).

97. *The Advantage of Spatial Deduction*

In Chapter 3 we compared the two deductive processes, the mathematical and the logical, from the present point of view. However, it might be useful to go into a bit more detail, given the importance of the question and the illusion of reality the spatial construct creates in us. Again let us say that we are obviously dealing in both cases with completely analogous things, namely, attempts to make reality rational — to make it seem to conform to the needs of our intellect or, if one wishes, to show that from this point of view it could not be other than it is, that it is what it is *necessarily*. Now this cannot be done except by recreating it entirely with the aid of notions drawn from within ourselves. This is exactly why we found spatial deduction so clearly preferable to logical deduction; the structures of spatial deduction do not have the obvious nature of hypostatization that so markedly characterizes the Hegelian concepts. This is in fact a prerogative of mathematical, and more particularly geometrical, deduction: the prerogative of revealing a very extensive agreement between our mind and reality. Scientists and philosophers, as we saw (§23), have used this agreement to demonstrate, according to their particular persuasions, either the reality of the external world or its dependence on idea. Here idea and reality actually seem to merge. Kant saw this quite clearly when he opposed "pure philosophy, or metaphysics," which is "pure rational cognition from mere concepts," to cognition that undertakes "the construction of concepts by means of the presentation of the object in *a priori* intuition" and "is called mathematics."[4]

98. *The Given in the Relativistic Deduction*

Undoubtedly geometry today no longer seems to us as strictly *a priori* as Kant, and certainly most mathematicians of his time, took it to be. It is, of course, accurate to say that the situation has not been essentially modified by the introduction of the new theory and that Euclidean geometry also contained just as many *givens* in its initial axioms and postulates (§86). However, it is nonetheless certain that, by the very fact that we no longer accept these axioms and postulates as they are furnished to us by our immediate perception — on the contrary, we reject them in favor of other axioms and postulates of which perception is, to say the least, completely unaware — we are much less inclined to claim that we have derived everything from our intellect alone. We know that at the very beginning of his career Kant attempted to deduce *a priori* the tridimensionality of space. Later he tacitly abandoned this demonstration, probably because he had realized it was unsound. But it is likely that he saw there only a temporary lacuna that would be remedied by later efforts. Schelling as well as Hegel, each in his own way, later attempted the same task, but succeeded only in making it even more clear that this property of space is an irreducible given (ES 2:143). We should not be too shocked at these fruitless attempts; let us simply note once again that what bothers us so much about them is, above all, the extramathematical way their authors proceed. For the relativist in his turn (as we pointed out above in §90) would have us "discern something of the reason why the world must of necessity be as we have described it," and, since this universe is only space, he therefore means to deduce why space has the structure it does and not another. Furthermore, Weyl believes he can support the claim that his theory "does not only lead to a deeper understanding of Maxwell's Theory but the fact that the world is four-dimensional, which had hitherto always been accepted as merely 'accidental', becomes intelligible through it" (STM 285).

Nevertheless, even the most confirmed relativist will not claim to have a direct intuition (in the Kantian sense) of the specific properties of his space.

99. *All Reasoning Begins with Perception*

Given this, and with a clear understanding that geometrical deduction could not claim to free itself completely of all outside elements, must one see here a radical distinction between this kind of deduction and deduction by [nonmathematical] concepts? Is it not clear, rather, that any concept must at bottom contain elements of this [empirical] origin; for how would reason form concepts if it had no sensations to set it in motion? According to Schelling,

the statement that the science of nature should be able to deduce all its propositions *a priori* has sometimes been understood to mean that science ought to dispense with experience entirely and draw its propositions from itself without the intervention of any experience. This claim is so absurd that even the objections brought against it merit nothing but contempt. Not only do we not know this or that particular thing, but we do not know anything, from the beginning and in general, except by experience or by its intermediary; consequently, everything we know is composed of propositions about experiences.[5]

But had not Plato, the eternal prototype of all idealists, already said that it was "impossible to reason without starting from a perception"?[6] Thus, from this point of view, there can be only a question of degree between these two kinds of reasoning, and the relativistic doctrine is indeed an *a priori* deduction of reality, just like Descartes's and Hegel's.

100. *The Tendency toward Idealism and Realistic Convictions*

It is precisely this kinship that can make us better understand, or rather understand from a more general point of view, why relativity theory's move in the direction of the abolition of the real is both necessary and (as Eddington's surprise demonstrates) unconscious. Although the Einsteinian physicist, like all physicists, is basically a realist, the very success of his deduction leads him to a structure that is just as basically idealistic. But this idealism, though quite obvious, is discovered only at the end. Where the relativist really joins paths with Hegel is in this abolition of reality, in acosmism (according to the somewhat pejorative expression devised by the German philosopher, although he of course meant to characterize thereby a philosophy he considered very different from his own). Moreover, it would be all the more pointless to try to attribute idealistic views to him before he has arrived at this final point, since he is not even conscious along the way of where he is heading. The very fact that Eddington was surprised proves to what extent even the physics that seems to come closest to idealism is nevertheless essentially realistic.

What allows it to remain realistic even while 'constructing' reality *a priori*
is, of course, the privileged character of the spatial noted above (§§25, 97) in
comparing Hegel's deduction to Descartes's. Also, the fact that the relativists
frequently tend to underline what is empirical in the foundations of their
geometry (§86) — although no deduction can dispense with elements of this
origin, as we just pointed out by means of the quotation from Schelling — is
only a manifestation of the dominant and contradictory concern not to allow
reality to disappear in the very process of demonstrating its rationality.

Thus, in relativistic physics as in physics in general, the tendency toward
idealism coexists side by side with realistic convictions. The former is attested
by the fact that the entire theory constitutes an attempt at a universal deduction of reality, while the latter find their expression in the limits encountered
by this explanation, which we shall now attempt to explain.

NOTES

[1] Schopenhauer, *Die Welt als Wille und Vorstellung* (Leipzig, 1877), 1:10.

[2] [The bracketed insertion is Meyerson's.] Weyl, for his part, stresses the fact that the
electron (which, as we know, is supposed to be the ultimate constituent of matter),
"formerly regarded as a body of foreign substance in the non-material electromagnetic
field," is now seen as only a very small, not very well-defined region in which the electrical field and density have enormously high values. "An 'energy-knot' of this type
propagates itself in empty space in a manner no different from that in which a water-
wave advances over the surface of the sea; there is no 'one and the same substance' of
which the electron is composed at all times." Thus "it is not the field that requires
matter as its carrier in order to be able to exist itself, but *matter* is, on the contrary, *an
offspring of the field*" (STM 202–203). (We have allowed ourself to modify the translation slightly in order to follow the original a bit more closely [here the Brose translation
(STM) matches Meyerson's text unchanged]. Concerning the relations between the field
and matter, cf. also p. 206 ff. on Mie's theory.) Thus matter literally dissolves in the
electrical field, which, as we know, is only a function of space, because there remains,
according to Mie's formula, only one "unique universe-substance," the same substance in
empty regions as in those occupied by matter (Gustav Mie, *La théorie einsteinienne de la
gravitation*, trans. Rossignol, Paris, 1922; cf. Weyl's comments on this theory, STM 200).

[3] Eduard von Hartmann, *Das Grundproblem der Erkenntnistheorie* (Leipzig, 1882),
p. 22.

[4] Kant, *Metaphysical Foundations of Natural Science*, trans. James Ellington (Indianapolis: Bobbs-Merrill, 1970), p. 5 [Preussische Akademie ed., 4:469].

[5] Schelling, *Einleitung zu dem Entwurf eines Systems Naturphilosophie, Sammtliche
Werke*, 1st series (Stuttgart, 1858) 3:278.

[6] Plato, *The Republic*, 523–524 [Meyerson's citation seems to be in error here].
Aristotle similarly stated that "anyone who had no sensation would be incapable of
learning or understanding anything at all" (*On the Soul*, Bk. 3, Ch. 8, 432[a]6).

CHAPTER 12

ESSENCE AND EXISTENCE

101. *The Nature of this Distinction*

However, before investigating the limits imposed on the relativistic conception insofar as it is a theory arising from physical science, we must identify a limit it incurs insofar as it is a universal deduction of reality.

It is obvious that the relativist does not pretend to discover why a particular 'pucker' of space is found at one point and not at another; he accepts this as a fact because it is simply the translation of the observation we customarily speak of as the presence of a mass or an electrical field. He is content to assert that once this fact is given he understands how the gravitational and electrical phenomena follow. In other words, his explanation is limited to what concerns the *essence* of the phenomenon, leaving completely aside everything concerning its *existence*.

102. *Its Role in Medieval Thought*

This distinction goes back to the Arabic and Jewish philosophers. It plays a considerable role in the Christian philosophy of the Middle Ages. In the eleventh century, near the beginning of this philosophy, Saint Anselm makes use of this distinction as the point of departure for his famous proof of God's existence: God exists by his essence and owes nothing to his existence – this proof, it must be added, being applicable to God alone.[1] Later thinkers insist on this point. "In everything that is created," Saint Thomas states, "the essence differs from its being and stands in the same relationship to it as potentiality to its actuality." Similarly, Henry of Ghent declares that

things that are distinct from God can be considered in two ways, namely, first in that which concerns the being of their essence, which is the being of quiddity; secondly in that which concerns the being of existence.

Duns Scotus asserts that "the being of substance is twofold, namely, the being of essence and the being of existence."[2] These scholastic concepts recur in Bossuet's *Logic*: "In order to know a thing that actually exists, one must bring together two ideas: one of the thing in itself, according to its proper

101

essence, the other of its actual existence," because "in all things, with the exception of God, the idea of essence is distinct from the idea of existence."[3]

103. *Its Role in Modern Philosophy*

These concepts survive in Spinoza's philosophy, so much so that Bergson was able to say that "all of Spinozism rests on the distinction between essence and existence,"[4] and Leibniz also refers to it.[5] Later thinkers, however, do not stress it nearly as much. This is because the preoccupation with this kind of reasoning disappeared after the ontological proof of the existence of God was destroyed by Kant.[6] This is also why we felt compelled to pick out a few quotations, almost arbitrarily, from the enormous volume of reasoning the scholastic masters in particular devoted to this area, expending a wealth of subtlety in the process. These quotations tend to show to what extent this thought, so unfamiliar to us today, seemed fundamental to our predecessors. In any case, Schelling returns to the question and uses it to try to justify how, in his system (contrary to Hegel's), the mind, "starting from the abstract subject, from the subject in its abstraction," enters, "from the first step, into nature," so that "there is no need for subsequent explanation in order to make the transition from logic to reality," for the theory "is exclusively concerned with the how of things without pronouncing on their real existence." In fact, this existence must be considered due to chance, to the "surging up" of a "primitive spontaneity" that can be grasped only by experience. "Even in the case of something reason has grasped, only sensation [*der Sinn*] can teach us that this thing *is*, that is to say, that it exists."[7]

104. *Its Role in Relativism*

It will be readily understood, moreover, that the reappearance of this conception in Schelling is not a matter of chance. In affirming the identity of thought and being, and consequently the deducibility of the physical world, if only in principle, this philosopher is actually attempting to do the same thing the scholastics did when they intended to deduce the existence of God from the idea of God. But since for him there is no perfect and unique being involved, Schelling retains only that part of medieval thought dealing with the limits of the deduction of the concrete. By the same token, if we in our turn feel compelled to insist on this distinction, it is because relativism is engaged in the same sort of enterprise when it asserts (as Eddington put it in the passage quoted above, §90) that it intends to enable us to "discern something of the reason why the world must of necessity be as we have described it." Therefore, the restriction as formulated by Schelling is completely valid here.

Thus there is indeed in this case a limit that is not restricted to relativity theory alone, nor even to mathematical deduction in general, but is the limit of any *a priori* explanation whatsoever.

NOTES

[1] Cf. Alexandre Koyré, *L'idée de Dieu dans la philosophie de saint Anselme* (Paris, 1923), pp. 55, 222, 240.

[2] Saint Thomas, *Summa Theologiæ*, Pt. 1, qu. 54, art. 3, *Opera Omnia* (Rome, 1889), 5:47: *In omni ... creato essentia differt a suo esse, et comparatur ad ipsum sicut potentia ad actum.* Henri Goethals de Gand, *Quodlibeta* (Paris, 1518), CLVIII, O: *Ea quæ alia sunt a Deo, dupliciter possunt considerari: uno modo quoad esse essentiæ eorum, quod est esse eorum quidditativum, alio modo quoad esse existentiæ.* Duns Scotus, *Quaestiones super Analyticas Superiores*, ed. Waddington (Lyon, 1639), 1:392: *Substantiæ duplex est esse, sc. esse essentiæ et existentiæ.*

[3] Bossuet, *Logique*, Chs. 39, 40, *Œuvres complètes* (Bar-le-Duc, 1863), 12:51–52.

[4] Cf. Constant Bourquin, *Comment doivent écrire les philosophes* (Paris, 1923), p. 16.

[5] Leibniz, *Opuscules et fragments inédits*, ed. Couturat (Paris, 1903), p. 9.

[6] With all due respect to the learned author of the above-mentioned work on Saint Anselm, we see great merit in Kant's demonstration that – contrary to what had been affirmed not only by the Scholastics, but by Descartes and Malebranche as well – essence does not include existence in the case of God any more than in any other case. It is certain that the concept of perfection stressed by Koyré, following the thinkers of the Middle Ages, hardly seems to modern man to be able to support the weight of so heavy a deduction. The fact that Kant was already anticipated to a certain extent in the Middle Ages by Gaunilo does not detract from his achievement, especially since Gaunilo's objections apparently had remained without any effect on his contemporaries or his immediate posterity, while Kant, by the very fact that he shows synthetic reasoning to be admissible only in mathematics, has identified the true basis of the impossibility reason comes up against here.

[7] Schelling, *Zur Geschichte neuerer Philosophie, Sämmtliche Werke*, 1st Series (Stuttgart, 1861), 10:146 ff.; *Philosophie der Offenbarung, Sämmtliche Werke*, 2nd Series (Stuttgart, 1858), 3:58, 104 [Meyerson has inserted the bracketed German].

CHAPTER 13

DIVERSITY

105. *The Simplification Brought About by Relativism*

If we now turn to the strictly scientific aspect of the theory, the first thing we notice is that, in reducing reality to space, the relativistic doctrine necessarily ends up denying to a great extent the diversity of that which it is trying to explain. Thus we must look particularly to diversity in our search for the limits we have just discussed. This point needs further clarification, however.

Einstein's supporters stress the simplifications resulting from their conceptions. Anyone familiar with at least the rudiments of the theory knows that this is justified and that relativistic geometry really is simpler than the physics it replaces.

Indeed, it is clear that the mind could not be expected to accept alleged explanations of physical complexities that are just as complicated as what they are trying to explain and whose only advantage is that they are mathematical in nature. On the contrary, the theoretical representation must make good use of the virtue inherent in mathematical, and more particularly geometrical, deduction — what Henri Poincaré has called the "power of grouping." [1] Wien points out that if one uses a sufficiently complicated geometry, it will always be possible to represent any empirical observations whatsoever. He sees this as the failing of Weyl's theory, which, he says, "ends up transferring to geometry some of the difficulties inherent in a representation of the laws of nature" (WW 208; cf. 158). Needless to say, we lack the competence to enter into this debate. But we are convinced that the consensus of competent physicists will quickly settle the question: Weyl's theory or any analogous one will be able to prevail in science only if it is markedly simpler.

But no matter how well relativism in general has succeeded in this regard, it is certain *a priori* that it cannot eliminate all diversity; therefore, while physics becomes simpler, the geometry that replaces it must become more complex — not to the same degree, it must be granted, given the above-mentioned virtue inherent in the theory, but nevertheless to some degree. And we have seen that this is indeed the case. First of all (leaving aside the complications resulting from the inclusion of the dimension of time), instead of the absolute simplicity of Euclidean space we have Einsteinian space with its 'puckers'. And then, if we mean to include electrical phenomena in the theory, there are other even more formidable complications added by the theories of Weyl and Eddington. The ordinary man finds these complications incomprehensible or absurd, but the Einsteinian physicist, whose mind has been made flexible by familiarity with mathematical formalism, declares that he can perfectly well picture this reality that the ordinary mind so completely lacks the flexibility to imagine. For him then, every feature of diversity that can be reduced to the spatial, or rather to the hyperspatial, seems to be explained, part of it being abolished and the rest transferred to space.

106. *Where has Physical Reality Gone?*

Is that all there is to physical diversity?

Of course not, for if that were the case, relativity would have accomplished the essentially chimerical tour de force of deducing being from nothingness, since space, in spite of all the complications imposed on it, can never by itself be anything but nothingness. Something must therefore be left over, a realm where physical reality took refuge when geometry was not able to contain it.

107. *Irreversibility*

We have already touched upon this realm in speaking of the irreversibility of phenomena (§70) in connection with the spatialization of time required by relativity. In pre-Einsteinian mechanics, as we said, irreversibility seems to be an anomaly, because mechanics, as a 'rational' science, demands that things remain what they are and that, given the effect, the cause can be reproduced (§72). Consequently, if one wishes to approximate reality, one is forced to formulate a special principle. This is Carnot's principle, which constitutes a clear-cut expression of the so-called anomaly; for this very reason it was only belatedly and reluctantly accepted by the scientific community (§75). Admittedly, it has been possible to give this principle a mechanical interpretation by means of a device introducing statistical methods. However, there remains something of the original character of the observation underlying this principle, something incommensurable with reason, for one must allow a fundamentally improbably state "at the beginning of time," without trying to make this state arise (Svante Arrhenius to the contrary) from another more probable one (cf. ES 1:206 ff.; 2:405).

108. *Relativism Makes the Situation Worse*

It is clear that the theory of relativity in no way improves this situation; on the contrary, it tends to make it worse. This is because, while things remain more or less unchanged with respect to statistical explanation, the fact that one is attempting to assimilate the temporal dimension into the spatial dimensions underlines the lack of analogy between them. Why can we not move freely in time as we do in space, as would seem to be implied by the perfect *isotropy* Eddington attributes to the universe? We need a principle to guard us against any temptation we might have to go backward in time, to assure us that under no circumstances can we telegraph into the past. It is quite possible that a judicious use of statistics might here again make the anomaly less flagrant.[2] But it is nonetheless certain that the theory of relativity, taken by itself, has just the opposite effect. Furthermore, we have noted that relativists are not overly conscious of this difficulty, and their writings even tend to hide it to some degree, in obedience to a natural propensity of the human mind.

109. *Discontinuity*

On the other hand, the relativists, or at least some protagonists of the doctrine, are deeply concerned with another aspect of the eternal enigma of diversity, namely, the discontinuity of physical reality. We have seen how clearly Eddington understands that the principles underlying the relativistic doctrine are laws of the mind. But he assumes that, in addition to these laws, there are others that are "genuine laws in the external world," that is, propositions independent of the "despotism of the mind." "The most important of these, if not the only law, is a law of atomicity." Admittedly, the author does not seem altogether sure that the existence of such laws has been definitively established. On the contrary, he admits that if our analysis "could be pushed further to reach something still more fundamental [than relativistic point-events], then atomicity and the remaining laws of physics would be seen as identities" (STG 198—199),[3] that is, as still dependent on our mind. Let us hasten to add that we find these reservations unjustified. There *must* be something in matter that is not of our mind, that actually resists it and remains opaque — otherwise physical reality would dissolve into nothingness. Certainly this final impenetrable element does not force itself on our attention, but that is because we are not looking for it. What reason seeks to do is to take hold of reality, to see reality as entirely conforming to the fundamental laws of reason. This is clear in physics as elsewhere, and the history of the laws of conservation, insofar as they are opposed to Carnot's principle, is sufficiently eloquent testimony to this. It also explains why, as Eddington observes, "where physics has achieved its greatest success is in the discovery of laws which largely belong . . . to the category of subjective laws," that is to say, laws demonstrating the agreement between the mind and reality. On the other hand, "until now very little progress has been made on the rocky road" that leads to the discovery of laws "which have not their origin in the mind" and "over which it has had no control."[4] This is a very apt observation, and it is certainly to be expected that although rationalization — the affirmation of an agreement between our intellect and nature — becomes more exact as physics progresses, there will at the same time be an increasing number of observations concerning the limits of this agreement, attesting to the existence of something in reality that our reason is incapable of incorporating completely.

110. *Absolute Measures*

This is indeed what has happened in the last few years. Planck seems surprised at the "coincidence" that, just

at the time when the idea of general relativity is making headway and leading to unexpected results, Nature has revealed, at a point where it could be least foreseen, an absolute invariable unit, by means of which the magnitude of the action in a time space element can be represented by a definite number, devoid of ambiguity, thus eliminating the hitherto relative character (PRB 163; Eng. 112).

For us, as we have just seen, this is not at all a matter of chance.

The absolute to which Planck alludes is that posited by quantum theory. This conception is basically the same kind of affirmation of discontinuity as the atomic theory, but, unlike the atomic theory, of altogether recent origin. It would perhaps be useful to look more closely at the difference between the two theories.

111. *The Concept of the Atom*

It is well-known that the concept of the atom is a very ancient one, which originated in Greece and elsewhere at almost the same time as physics. Its role in physics is sometimes conspicuous, sometimes more in the background, depending on whether a particular stage of the evolution of this science brings forward more or less evidence of discontinuity. Finally, at the end of the last century, our belief in discontinuity is firmly established by a whole series of experimental results assigning a number and measurable dimensions to the particles of matter. However, as revealing as these discoveries are for the physicist, certainly exceeding his fondest hopes, basically they only confirm and make more precise a presupposition he has always held.

112. *Quanta*

Quanta, on the other hand, appeared quite suddenly. We spoke of this affair at length in our earlier work (ES 1:39 ff.) and therefore shall only recall here that it involves observations first made in connection with the phenomena of black body radiation. It was Henri Poincaré who clearly demonstrated that the experimental formulas deduced from the observation of these phenomena could be explained only by assuming that they represented discontinuous processes, but it remained for Planck to reduce the whole body of these anomalies into a system. Now this system posits the existence of a kind of discontinuity that had never been imagined up to that time – which is not at all surprising since this discontinuity, which is the discontinuity of motion, seems to contradict completely the very foundations of our conception of the nature of reality. This discovery, therefore, unlike the discoveries concerning atoms, was altogether disagreeable. Since then physicists have done all they could, and more, to try to relate these phenomena to other less paradoxical ones, in whatever small measure was possible. But this seems to have been completely fruitless. One of the strongest supporters of quantum theory, who, in a recent work, celebrated its "astonishing successes" and expressed his admiration for the courage of the few clear thinkers who defended the necessity of the new doctrine and "fought against tradition," feels compelled to recognize that we "have not come one step nearer to understanding the heart" of phenomena, given that one still does not know "where . . . the deeper cause lie [s], which brings about this discontinuity in nature." Thus he concludes that "we are still completely in the dark."[5] Furthermore, it seems likely that the darkness will grow deeper, or at the very least more widespread. As a matter of fact, since these anomalies have come to the physicists' attention, scientists are discovering them just about everywhere. For lack of a better solution, physicists are at least happy to be able to connect them more or less easily to the anomaly observed by Planck. In this manner quantum theory has successively made inroads into the

molecular theory of solid bodies, the theory of gases and the theory of the
constitution of the atom.[6]

113. *Relativism and Quanta*

From the standpoint of relativity theory, quantum theory, just like the
affirmation of the existence of atoms, certainly appears to be something
absolutely distinct and unassimilable. The relativists themselves are quite
aware of this; they feel that their explanation encounters a serious obstacle
here — we might even sharpen this image by saying that the collision is painful. Einstein, for instance, believes that "the facts comprised in the quantum
theory may set bounds to the field theory beyond which it cannot pass"
(ER 23). For Weyl, "phenomena that we are with varying success seeking
to explain by means of the quantum of action, are throwing their shadows
over the sphere of physical knowledge, threatening no one knows what new
revolution" (STM 212).

114. *The Physical Resists Reduction*

Therefore, what is revealed by discontinuity, whether of atoms or quanta, is,
as in the case of Carnot's principle, the component of reality that cannot be
deduced from the fundamental principles of reason. Moreover, since in the
theory of relativity the rational appears to be purely spatial — geometrical —
in nature, one can say that what prevents us from making everything rational
in this case is the properly physical, as opposed to mathematical, component
of reality.

Obviously a great advantage of the theory of relativity is how clearly the
scientific data it is unable to encompass stand out as soon as one considers
this problem. In this respect, the theory itself marks its own limits, so to
speak. And one also sees why this is so: it is because we are confronted here
with a far-reaching attempt at universal deduction pursued with extraordinary
daring and vigor, starting from a base that is at once precise and narrow.

In considering the particular form that the deduction has assumed in the
present case, we can say that the *mathematization* of nature, by the very fact
that it consists in making nature rational, must inevitably end up by exposing
what is irrational in it. As Brunschvicg says so well, "the mathematical form,
by its very nature, reveals the given that cannot be reduced to this form, i.e.,
that which is specifically physical" (EH 407).

NOTES

[1] [*"Vertu du groupement."* Although we have not found a passage in which Poincaré actually uses this phrase, he is probably referring to the insight into the fundamental properties of geometry which he feels can be achieved through the application of the mathematical notion of a group. See his 'On the Foundations of Geometry,' *The Monist* 9 (1898): 9ff., or *La science et l'hypothèse* (Paris: Flammarion, 1923) p. 83 (*Science and Hypothesis*, trans. George Bruce Halsted, New York: The Science Press, 1905, p. 49). We are indebted to Arthur Miller of the University of Lowell and Harvard for calling the *Monist* article to our attention.]

[2] It should be noted that Hermann Weyl, who believes moreover that *"statistics* plays a part in [the solution of these problems] which is fundamentally necessary," believes at the same time that there must be a relation between the irreversibility of phenomena and the existence of two kinds of electricity: "This connects the inequality of positive and negative electricity with the inequality of Past and Future" (STM 311 [the bracketed phrase is included in Meyerson's quotation, but appears only as "it" in STM]).

[3] [The bracketed insertion is Meyerson's.]

[4] Eddington, *Espace, temps et gravitation*, trans. J. Rossignol (Paris, 1921), pp. 245–246. [We have translated the first two quotes directly from material added to the Rossignol translation and not found in the English original (see also Ch. 20, n. 3 below). The remaining two short quotations are included in STG 200.]

[5] Fritz Reiche, *Die Quantentheorie, ihr Ursprung und ihre Entwicklung* (Berlin, 1921), p. 160 [*The Quantum Theory*, trans. H. S. Hatfield and Henry L. Brose (London: Methuen, 1922), p. 125]. Cf. Ch. 16, n. 7 below.

[6] Reiche, pp. 37, 88 [Eng. 29, 68]. We shall deal later (§ 127) with the manner in which the concept of quanta was introduced into Rutherford's theory of the atom after Bohr's work.

CHAPTER 14

INTERPRETATION

115. *Abstract Number and Concrete Magnitude*

It must be noted, however, that it is not nearly so clear in the actual work of scientists that they exclude everything relativism, as a physical theory, regards as nondeducible. This is, in fact, an important limitation, which we believe is often overlooked.

Mathematical deduction, *calculation*, can quite obviously furnish only abstract numbers; everything the least bit *concrete* in any way whatsoever must come from somewhere else. We discussed this in Chapter 3 (§ 28), where we pointed out — citing Aristotle, Kant, and Borel — that an element that is not purely rational can be discerned in the geometrical itself. This is all the more true as soon as we leave the field of pure mathematics for physics. "The equations of physics," says Lippmann, "are not analytical relations; they are quantitative relations between qualitatively irreducible magnitudes."[1] One need only examine the relativistic deductions to see this observation fully verified. The relativist himself does not seem to be particularly conscious of this; he is a physicist and as such thinks primarily about concrete reality, so that interpretation of the results of his calculations seems natural to him. When he transforms abstract number into concrete magnitude, he has the illusion that he has added nothing. The only reason he can do his calculations is because he first stripped his datum of any qualitative character and considered it as an abstract number. If he had retained the true nature of his most elementary numerical data — indications of length and distance — he would not even have been able to multiply them to obtain areas, since *a priori* one cannot any more multiply meters by different meters than one can multiply meters of fabric by liters of milk. Moreover, having made this initial abstraction, unless he wished to see his results lose all meaning, he had to undertake the inverse operation at the end: he had to attach something — a qualifier — to the number that is the strict result of his operation in order to transform it into a *coefficient* (cf. ES 2:209 ff.). For he needs coefficients in order to deal with the physical world, and these coefficients, these concrete magnitudes, can perform their assigned task only because they no longer have the nature of pure numbers. In other words, they no longer belong to the class of scientifically unreal concepts that are, in Hartmann's words, fundamentally unsuited for reconstituting reality; they already contain an element of reality, obviously a qualitative element.

116. *The Relativist's Illusion*

But the relativist is so happy and proud (quite rightly so, we might add) to have succeeded in transforming the physical world — in *sublimating* it in such a way that the product of this operation seems to coincide with what the mathematician has obtained elsewhere by purely rational procedures — that he is apt to lose sight of the fact that this agreement is not and *cannot* be complete. Calculation could only provide a purely numerical result and, in order to transform it into a physical principle, interpretation must enter in. Thus, to take a particular example, it is not accurate, strictly speaking, to say, as we do, that Maxwell's equations are only the simplest invariant form of the relation between four parameters. For Maxwell's equations are physical laws; the invariant form that we deduce constitutes only a purely mathematical framework for them, and if a mathematician had deduced these formulas as an exercise in pure calculation, he would have been quite incapable of saying what they meant.

117. *Even the Relativistic Concept of the Spatial is the Result of Interpretation*

Thus the entire apparatus of the Einsteinian deduction needs interpretaton in order to produce reality. Even the idea of the spatial as it is found in this conception is not a product of the deductive process alone. It results from the fact that we started from a spatial image and then, when this was modified by going beyond the limits of the concept of space provided by naive perception — for example by adding new dimensions to the three of Euclidean geometry — we more or less consciously endowed what was added with attributes that are exactly those of a spatial dimension (cf. § 120 below).

Because of this, and because mathematical formulas by themselves are so ill-suited to showing us how we should use them, it has been possible to use a considerable part of the collection of symbols constituting Einstein's theory (at least the special theory of relativity) to create an entirely different construct. This is what A. A. Robb, in particular, has attempted in his quite remarkable works.[2] Although we cannot speak concerning the mathematical soundness of these works or their value as frameworks for physical deduction, even the uninitiated cannot fail to be struck by the extent to which the meaning of Einstein's formulas has been modified by interpretation.

NOTES

[1] Gabriel Lippmann, 'La théorie cinétique des gaz et le principe de Carnot,' *Congrès de physique de 1900*, 1:550.

[2] A. A. Robb, *Optical Geometry of Motion* (Cambridge, 1911); *A Theory of Time and Space* (Cambridge, 1914); and *The Absolute Relations of Time and Space* (Cambridge, 1921). We are only familiar with the last of these.

CHAPTER 15

THE RELATIVISTIC IMAGINATION

118. *The Existence of a Limit*

We have already called attention on several occasions to the fundamental observation that relativism, in substituting a different reality for sensible reality, is therefore obliged to some extent to call upon the imagination to represent this new reality. This constitutes a further limitation of the theory, one that has been little noted (at least if one discounts the protests of opponents who reject the theory out of hand because they claim that they are unable to imagine what it postulates). This limit has escaped notice because it is by nature imprecise and, moreover, as we are about to see, can never be fixed, given that we are unable to predict ahead of time what we may or may not be able to imagine under circumstances that are not fully actualized at the moment we make our judgment.

It is nevertheless clear that there does exist a limit of this kind. No matter how convinced one may be of the parallelism between geometry and arithmetic, and no matter how attached one may be to the idea of pushing this analogy as far as possible, one would certainly hesitate to allow a fractional number of dimensions. And it would obviously be easy to find example after example showing how absurd it is to transpose into geometry *all* the complexities that abound in analysis.

119. *The Imaginary Quantity in Algebra*

However, this observation raises a question. One can ask whether the relativistic theories currently being presented do not already go too far in this respect and do not exceed the limit of the *imaginable*. Whatever purely abstract beauty the conception might otherwise have, can the imagination encompass it, and if not, will the theory really have any significant explanatory value, even for the initiate?

One of the first things we must deal with in this context has to do with the imaginary component of the temporal dimension. There is no doubt that in algebra the *imaginary* is opposed to the *real*. Is it not, therefore, contradictory to admit that the relativist is able to accomplish the feat of imagining the real existence of this temporal dimension?

114

Let us recall, however, how very clearly the relativists affirm the existence of their universe. Their affirmations obviously prove that they are not struck by the contradiction we have just pointed out. Furthermore, we need only examine things a little more closely to see how this has come about.

The imaginary quantity in algebra was indeed originally understood in the strict sense of the definition mentioned above. But since that time, constant use has made the concept more and more familiar to the mathematician, and by that very fact it has come to seem less strange, less chimerical.[1] Moreover, since Argand we have become accustomed to a *geometrical* representation of this imaginary quantity, that is, a representation by figures placed (it goes without saying) in real space. And anyone at all familiar with treatises on relativity knows that this is how they proceed when they set out to deduce that which involves the temporal dimension. In particular, there is constant reference to the *conical, hyperbolic* or *cylindrical* spatio-temporal universe, and figures completely analogous to those of ordinary geometry are drawn in it. It is only later, after finishing the deduction, that the correction is made by reestablishing the imaginary nature of the dimension. At that time one is obviously forced to abandon the image, but one is hardly aware of doing so, and it remains no less true that throughout the deductive process the imagination was dominated by this representation. Thus the contradiction tends to disappear; nevertheless it teaches us that it is not easy to define the limits in question here.

120. *Spatial Image and Algebraic Formula*

Furthermore, one must admit that authoritative versions of relativity theory do not enlighten us on this point as much as one might have wished. Certainly these theories are essentially realistic, and necessarily so, as their proponents are most often well aware. But although the works as a whole are undeniably imbued with such a spirit, lapses can sometimes be observed in the details of the deductions: one speaks as if it were a question not of something real, but of exclusively mathematical expressions. We need hardly add that these anomalies are particularly likely to occur when the complexity of the mental image threatens to become too formidable. Thus Eddington does not mean for us to add another dimension to the ones with which we are familiar, that is, to imagine four-dimensional space floating in a space of five dimensions. "The value of the picture to us," he says, "is that it enables us to describe important properties with common terms like *'pucker'* and *'curvature'* instead of technical terms like 'differential invariant' (STG 85).[2] Clearly, what this physicist is treating so lightly here is in fact of capital importance. This is amply demonstrated by the fact that Eddington himself speaks constantly of these *puckers* and *curvatures* throughout his book; it would undoubtedly be very difficult for him to do otherwise without seeing the theories he so ably and forcefully sets out lose the very quality that makes them good explanations. After all, the explanatory power of these theories derives from the fact that we are dealing with a true concrete geometry, not with theorems of abstract analysis.

On this point let us also go back to Weyl's statement quoted in § 54, where he says that "the two essences, space and time, entering into our intuition have no place in the world constructed by mathematical physics." There is a strong analogy between the attitude indicated by these words and the one that inspired Eddington's statement. It seems clear to us, however, that Weyl's statement, taken literally, would contradict the very foundations of the relativistic doctrine, which, as the reader has been able to see, is most certainly a spatial construction of reality. Moreover, it would also contradict the necessary foundations of physical explanation in general as we have described them in Chapter 3. As a matter of fact, from the picture of the world presented by immediate perception — which can be considered as a framework consisting of space and time and a content that fills it — physical science has always endeavored to eliminate the content, conserving only the framework. Relativism, by resolutely eliminating all reference to properties peculiar to time, retains only those peculiar to space. If we were to eliminate them as well, there would be nothing left to give us even the illusion of a reality, however pure it might be; it is certainly impossible — as Aristotle amply demonstrated — to recreate the universe from pure quantity. Thus the relativists do not really try to do so, and it is easy to see that all their constructions are essentially spatial. As Weyl himself stated elsewhere, "in order to arrive at the idea of the metric continuum" they merely "enlarge the Euclidean idea of space." In other words, by proceeding in this fashion, they retain those properties peculiar to the notion of space, namely, qualitative properties (and this is consistent with what we suggested in § 117). Moreover, it is only these properties that allow them to arrive at a construction that is in any way acceptable to the imagination.[3]

It is significant that these two eminent scientists could make such a mistake and apparently think, even if only in passing, that it would be possible to have a concept of reality completely devoid of any element deriving from the notion of the spatial. This can be explained in part by the difficulties inherent in the kind of spatial representation relativism makes necessary; because of these difficulties the mind is sometimes apt to wish it could free itself from this necessity. Another more profound cause is at work here, however, as we shall see below (§ 158).

Furthermore, it is no accident that these statements were both made by physicists trying to extend the limits of relativity. Indeed, as long as we remain in the domain of Einstein's theory, strictly speaking, the situation is not yet too shocking, at least from the standpoint of the mathematician's imagination. More than half a century ago Helmholtz had already claimed that he was perfectly capable of picturing a curvature of space such as would become possible if one assumed an additional dimension, and Eddington speaks of the approaching image offered us by the convex mirror of a shiny doorknob. But if we then enter into Weyl's domain, will not even the freest and best-intentioned mathematical imagination lose heart and fall back?

121. *Poincaré's Prediction*

In the above discussion we have taken pains to express ourself cautiously; indeed, one cannot be too careful where affirmations (or negations) of this kind are concerned.

In the case of the imaginary character of the temporal dimension, we have just seen how easy it would be to err if one claimed to define the limits of the mathematical imagination without carefully examining all the circumstances, even when one is dealing with conditions that are fully understood and fully achieved at the present moment. But the task becomes immeasurably more difficult when one attempts to deal with the future. One is claiming in such a case to know ahead of time how the mind will behave under future conditions – conditions, moreover, that must be indeterminate, since it is impossible to know what future experimental observations will be or which of our present theoretical concepts will be shaken by them. As we have argued many times in our previous works, and as we believe has been established, we are totally incapable of reasoning effectively in the face of such vagueness, merely tentatively, so to speak, because our mind demonstrates its real capacities only when it applies itself to a problem with all the conviction and energy of which it is capable. The advent of the theory of relativity offers us a flagrant example of just the kind of errors likely to be committed in such a situation. Among the great minds that have distinguished our epoch, both in mathematics and in mathematical physics, certainly none has carried more authority, nor merited it more, than Henri Poincaré. His efforts in the epistemological field are also acknowledged to be serious and fruitful. Nevertheless, in speaking of the assumptions of Lobachevsky, Riemann, and Helmholtz, etc., who suggested that astronomical observations could lead us to recognize that our space is non-Euclidean, Poincaré urged that if we discovered anomalies of the kind they foresaw, we would certainly attribute them to the nature of light and not to the nature of space.[4] He was predicting how the mind would respond to conditions that could, of course, be roughly described but could not be known exactly; in any event, they were not yet actualized at that time. And as a consequence Poincaré was completely wrong. When the time actually came, human reason went in the direction he thought he could disallow.

Perhaps a parenthetical remark would not be out of place here. We have just been speaking as if Einstein's theory had been definitively established. The reader knows from our preface, however, that this is not at all what we believe. We can even use the concrete example just cited to clarify the earlier statement. Indeed, if physicists were one day, even in the very near future, to abandon relativity and return to the traditional concept of time and space (as a result of observations contradicting those put forth by supporters of the theory), the fact would nonetheless remain that at one point in its history scientific thought took a direction Poincaré thought impracticable.

122. *The Indeterminateness of the Limit*

This leads us to the conclusion that the limit of the mathematical imagination necessarily remains completely indeterminate. It can only be established by the prevailing opinion, not of educated men in general, nor even of scientists, but of the mathematicians and physicists of a given epoch. Moreover, it can only be drawn temporarily, since, as we have seen, it is variable in nature, depending on the acquired knowledge of that epoch and also, to be sure, on its habitual mental attitudes.

NOTES

[1] This was noted by Henri Poincaré well before the epoch that marked the advent of relativism. "Those who invented imaginaries," says the great mathematician, "hardly suspected the advantage which would be obtained from them for the study of the real world; of this the name given them is proof sufficient" (*La valeur de la science*, Paris, s.d., p. 143 [*The Value of Science*, trans. Bruce Halsted (New York: Dover, 1958), p. 78]. General Vouillemin observes that at the present time "the symbol in question, although unfortunately christened *imaginary*, no longer surprises the mathematician any

more than the symbol for a fraction, for example" (*Introduction à la théorie d'Einstein*, Paris, 1922, p. 155).

[2] [Emphasis Meyerson's.] Cf. also Weyl's statement in § 158 below.

[3] Henri Marais similarly explains that, instead of following a "strictly logical method" to arrive at "the idea of any continuous multiplicity at all," he will merely "generalize the properties of nonlinear multiplicities contained in a Euclidean space." Perhaps one does not arrive in this fashion at "the most general idea of space possible, but this disadvantage is offset by the fact that the arguments refer more or less directly to familiar geometrical images and remain in contact with what may be called 'geometrical experiences'." He adds: "Moreover, the theory of relativity uses only spatial notions that can be reached in this way" (*Introduction géométrique à l'étude de la relativité*, Paris, 1923, p. 97). We saw above (§ § 21 and 47) that Marais is well aware of the realistic bias of relativism.

[4] Henri Poincaré, 'Les géométries non-euclidiennes,' *Revue générale des sciences* 2 (1891) 774. Poincaré came back to this conclusion several times, which proves how sure of it he was. Cf. his *La science et l'hypothèse* (Paris: Flammarion, s.d.), p. 93 [*Science and Hypothesis*, trans. W. J. Greenstreet (New York: Dover, 1952), pp. 72–73], and *La valeur de la science*, p. 109. Poincaré's argument had been anticipated by Lotze; cf. Bertrand Russell, *An Essay on the Foundations of Geometry* (Cambridge, 1897), pp. 99–100 (§ 92). The discrepancy between Poincaré's argument and the fundamental assumptions of the theory of relativity was pointed out by Eddington (STG 9–10).

CHAPTER 16

THE APPEAL OF RELATIVISM

123. *The Initiates*

Thus relativistic reality must above all be imagined. Considered from this point of view, the difference between the physicist who accepts Einstein's theory and the ordinary man who finds it incomprehensible and absurd is that the former thinks he can imagine a reality that is beyond the latter's power of imagination (§105). But it cannot be denied that, however well the mind may have been prepared, it is always a painful operation to substitute a new reality for the one to which we are accustomed. This is because our imagination obviously tends to depart as little as possible from the norms presented by common sense perception and to retain as many features as possible of the common sense representation in the reality it posits. Here the distance between the two representations of reality is truly enormous, so much so that only very few minds seem capable of making the correspondingly prodigious effort. Is what might be called the relativistic faith therefore destined to be confined to such a narrow group of initiates?

124. *The Attraction of Hegelianism*

In an attempt to clarify this point, we shall once again make use of a comparison with Hegelianism.

Anyone at all familiar with the history of philosophy since the beginning of the nineteenth century knows how enormously successful Hegel's system was and how much it attracted the most vigorous and even the most level-headed thinkers in the most diverse areas of intellectual activity. This attraction was powerful and — despite what may have been said — durable, since we still find it today, notably in the Anglo-American Neo-Hegelian school, in Croce and other well-known Italian thinkers, and even in France, at least to some extent, in the late Hamelin and his followers. Elsewhere, in discussing Hegel, we attempted to identify some of the factors contributing to this appeal. But the principal factor was undoubtedly the very fact that the system boldly claimed to be an attempt at a complete deduction of all that constitutes human knowledge in all its domains. Such an undertaking — the

120

demonstration that everything coming to us from outside ourselves is never-theless comprehensible by our reason and in conformity with its demands — is something to which we unceasingly aspire with all the fibers of our intel-lectual being. No doubt we have at the same time a vague feeling that this desire is probably chimerical, that fundamentally the world is not meant to be, and cannot be, entirely intelligible. But this impression is repugnant to us; we suppress it, so to speak, and would be quite happy to see it contradicted. This is what Hegel claimed to have accomplished. He was mistaken — but we make this judgment from a perspective that is completely different from that of his contemporaries. If, on the contrary, we consider the mentality of men who had been prepared by Schelling and Fichte, indeed even earlier by Kant, to think of reality as being entirely dependent on the consciousness, and if we also appreciate the truly extraordinary boldness and vigor of mind that Hegel brought to his undertaking, his success is less surprising to us.

Had all those who believed in Hegel actually understood his system, or had they only grasped the rudiments of the deductions by which he started from thought and arrived at reality? One can boldly affirm that they did neither. The vast majority, certainly, were (to use Stein's apt expression) tormenting their brains to no avail in attempting to understand the enigmas of the *Phenomenology*. They believed, however, and Victor Cousin, who admitted that he himself had not "understood much" in Hegel's demonstrations, yet admired him greatly, tells us why. It is because Hegel was dogmatic, and consequently a doctrine of this kind, which promised to attain reality starting from a single base, thus accomplishing what philosophers had long seemed to promise in vain, exerted a truly irresistible attraction (cf. ES 2:107—108). Similarly, quite distinguished thinkers counted themselves among the sup-porters of the German philosopher, seduced by this or that brilliant applica-tion to a particular domain of the sciences of the mind, despite the fact that for them the foundations of the doctrine remained enshrouded in an im-penetrable fog.

125. *Comparison with Relativism*

From this point of view, the circumstances surrounding relativity theory seem to be so different that at first one would readily believe any comparison impossible. It is indeed certain that public opinion, far from being carried away by the same kind of enthusiasm that marked the apogee of Hegelianism in Germany (and elsewhere), feels rather disoriented by the physicists' claims. To avoid appearing too extravagant, the physicists constantly find themselves compelled to stress the experimental confirmations the theory has received. Despite these assertions and all the explanations attempting to make the concepts more acceptable to popular taste, people certainly tend to consider the entire system fantastic, even fundamentally absurd, because of their inability to understand it.

This is because, in the first place, all the brilliant superstructure that contributed so much to the success of the Hegelian system is completely lacking here: relativity has to do exclusively with physics and, furthermore, with physical phenomena so subtle they are unlikely to attract the attention of the public at large. In the second place, mathematical demonstration, which is the only kind used by the theory of relativity, has the immense advantage of being clearer and more precise than the (extramathematical) demonstration by concepts used by Hegel: the immediate import of a mathematical formula can never be as uncertain as that of many verbal formulas. But here the advantage is turned into a disadvantage. For the imprecision of words is eminently well-suited to the creation of illusions, while the naked clarity of symbols tends to dissipate them. It is incomparably more difficult for someone ignorant of tensor theory to convince himself he has understood one of Einstein's demonstrations than it was to be under the illusion one understood the way *Dasein* was derived from *Sein*.

Finally, although the new doctrine quite clearly aims at universal deduction, and moreover clearly acknowledges this fact, it does not, as we have seen, proclaim this as emphatically as Hegel did. Rather than flaunting their audacity, the relativists would certainly be inclined to apologize for it. And this lack of assurance is not calculated to fire the imagination.

126. *The Conviction Created by the Deduction*

If, however, instead of considering these things from the point of view of the public at large, one examines the attitude of the initiates, the differences we have pointed out tend to diminish greatly and, indeed, even to disappear. Does everyone who has accepted Einstein's concepts really imagine the *puckered* space of the general theory of relativity, and does everyone who follows Weyl and Eddington have a clear idea of the still more complicated space these physicists need to assume? It would appear that one can, without fear of contradiction, reply in the negative. The great majority of adherents surely grasp the mathematical symbols only from the outside, so to speak. They see that if they accept the fundamental principle, this abstraction yields concrete phenomena. And it is the fact of this deduction that creates their conviction, that makes them accept the bases on which the whole construct is built and set aside any doubt prompted by their faltering imagination. Thus their mental attitude strongly resembles that of Victor Cousin.

127. *The Physicists and Bohr's Theory*

This assertion can be verified by examining other areas of science. A quite recent example is found in Reiche's fine book on *Quantum Theory* mentioned above (§112), where he discusses Niels Bohr's hypothesis on the constitution of the atom. He begins by pointing out that the fundamental assumptions upon which it is built are very "bold and unorthodox." After examining the difficulties presented by these assumptions and noting that they cannot be reconciled with certain precise observations, he comes to the conclusion that this model [of the hydrogen molecule (H_2)] yields "no agreement between theory and observation" and that "the arrangement of the two nuclei and electrons must plainly be quite different."[1] But these pages are separated by others where the theory is laid out in detail and where the author stresses the agreement of Bohr's theory with a great number of facts. Its successes have been "surprising." A particular coincidence (the connection between Rydberg's number and universal constants) is "striking" and "forms a strong argument for its innate power." The agreement between Bohr's predictions and the results of spectroscopic experiments carried out over a period of years is "perfect." By grafting Sommerfeld's hypothesis onto Bohr's, replacing Bohr's circles with ellipses and introducing considerations based on the theory of relativity, one was able to explain all the spectral lines of helium "almost without an exception" and "thus proved strikingly the existence of the stationary paths of the electron and its relativistic change of mass." These circumstances, needless to say, depend in the strictest sense on the image that forms the foundation of Bohr's theory and could not be conceived outside the framework it offers. Similarly, considerations concerning atomic collision show "with convincing clearness that the *Bohr* conceptions have laid bare the nature of the construction and the mode of action of the atom with unprecedented lucidity."[2]

Moreover, Reiche's attitude toward Bohr's theory is far from exceptional; on the contrary, one might venture to say that it is typical. For example, it can be seen in Wien's recent book, which we have had occasion to cite several times. Apart from its other virtues, this work is particularly valuable because the distinguished author has brought together, without modification, writings from different periods of his life, thereby enabling the reader to follow the evolution of his thought. Thus, in 1905, this physicist points out that the energy radiating from an electron revolving around a center could only come from its energy of motion. It follows that the energy of motion must diminish, that is to say that the satellite (in this case, the electron) must get closer and closer to the center that attracts it and eventually fall into it. In short, the electron could not remain in its orbit very long, which is why "it is impossible to explain spectral lines by the radiation of planetary electrons." This is not a casual observation. On the contrary, Wien returns to these arguments time after time, showing in particular that Bohr's theory is internally inconsistent, given that, although it is based on electromagnetic laws, some parts of the theory contradict these laws. Thus, one must not lose sight of the fact that this concept, "in spite of its extraordinary success, contains logical gaps that cannot be filled, at least not without completely overthrowing the theoretical foundations" of the hypothesis. Later, however, these objections become weaker and weaker until the author finally comes to admit that

of all the atomic models it is Bohr's above all that has been confirmed experimentally, since this concept has enabled physics to construct a theory of spectral lines – a theory at the same time very simple and very complete.

As to his former objections, Wien does not forget them; he simply observes that the theory shows us that within the atom there must be "laws of an unknown nature that are essentially different from anything we have known up to now."[3]

Analogous statements are to be found in Planck. This physicist observes that the assumptions upon which Bohr's hypothesis is based can only seem monstrous to any theorist accustomed to "classical" concepts; hardly a generation ago any physicist would certainly have rejected it out of hand. But, he adds, "figures are decisive." And elsewhere he praises Sommerfeld's "elegant formula," which he deems a discovery of the same order as Le Verrier's calculation of the trajectory of the planet Neptune before any human eye had perceived it. After declaring that "we might be inclined to consider all these ideas [offered by Bohr's theory] as the play of a vivid but empty imagination," he nonetheless concludes that in view of the numerous experimental confirmations, "one cannot escape the impression that science has once again succeeded in uncovering to some extent the secrets of nature."[4]

Finally, a similar attitude may be seen in Sir Ernest Rutherford, who is, as we mentioned (Ch. 13, note 3), the author of the atomic model from which Bohr's is derived. In his presidential address before the British Association, he obviously accepts this hypothesis as an established fact, yet at the same time he is forced to admit that

we cannot explain why these orbits [of the electrons revolving in ellipses around the central focus] are alone permissible under normal conditions, or understand the mechanism by which radiation is emitted.

He consoles himself with the reflection that it is possible that "the atomic processes involved may be so fundamental that a complete understanding may be denied us."[5]

128. *The Positivistic Explanation of this Attitude*

We beg the reader's indulgence for the numerous citations we have just made. We are dealing here with a particularly marked manifestation of a state of mind quite common among physicists, and it is important to see it with complete clarity in order to grasp its full significance. The situation is actually much too obvious in itself to have entirely escaped notice. It is usually interpreted, however, as demonstrating that in some sense the physicist does not take his theories altogether seriously. Contrary to what he seems to affirm, he does not seek to know what is behind phenomena. If he is not embarrassed by the contradictions his assumptions clearly entail, it is because the models he constructs, far from being a representation of reality, are for him only a more or less convenient framework intended to connect the diverse observations that have been made and prepare the way for the discovery of others of the same sort. Thus Whitehead, in a book full of originality in which he develops a theory of relativity essentially different from Einstein's, seems to believe that if a scientist disregards the beauty and generality of a theory, it is because he is only seeking to draw from it "precise application to a variety of particular circumstances so as to determine the exact phenomena which should be then observed."[6]

129. *The Inadequacy of this Explanation*

Though not completely erroneous, this positivistic explanation, like others of the same sort, is obviously inadequate. No doubt the physicist does hold the usefulness of his theories in high esteem, both from the point of view of finding the proper relationship between established laws and with an eye toward the discovery of new laws, but this is not all he appreciates in them, nor even the main thing. The most important thing is the understanding of the phenomenon provided, or at least promised, by these theories. To return to the specific case of Bohr's theory, not only Reiche, but Wien and Planck as well, obviously could not be more convinced that they are dealing with reality and insight into reality (their statements on this point prove it, and in the case of the latter two, the reader will also recall our citations in §22). It will, therefore, be necessary to look elsewhere for an explanation of the obvious paradox in this way of reasoning.

130. *The Appeal of Rational Explanation*

It will be recalled that we came to see in Chapter 2 that science could not get along without a body of hypotheses concerning the true nature of things, and that the progress of science consists essentially in making this reality more and more rational. As a result, once it has made its choice — that is to say, has adopted the best hypothesis, the one that seems most rational — science unconsciously tends to put aside and somehow forget anything imperfect remaining in this concept, anything that does not conform to the canons of reason.

We might put this a little differently to make clearer what this state of mind has in common with that of the Hegelians. We might say that, since science has an imperious need of a reality and at the same time seeks to demonstrate that this reality is deducible, it is necessarily led to exaggerate the value of what it has achieved in this respect. This brings us back again to Victor Cousin's attitude toward Hegel's system. In both cases, indeed, we see the same cause at work: the appeal of deduction, the satisfaction the mind derives from demonstrating that reality conforms to reason. We have the vague but persistent feeling that this satisfaction can never be complete. Consequently we treat any systematic attempt to make things rational with an indulgence that is almost inconceivable, or at least rather paradoxical at first sight. At the same time we cling tenaciously to the indestructible hope that all the gaps in the representation of reality we have adopted will later be filled, that the contradictions will prove to be merely apparent and will disappear in a higher synthesis. This is the same state of mind that makes us attribute special importance to any fragments of deduction, no matter how disconnected they may seem and no matter how improbable their basic assumptions. An altogether convincing example of this is provided not only by Bohr's theory but by quantum theory as a whole. The image serving as a framework for quantum theory is flagrantly improbable and affronts all the norms of our understanding.[7] It is based on a fundamental discontinuity in the motion of elementary particles, which are assumed to be able to move at certain velocities that are multiples of each other but not at intermediate velocities and to be able to describe certain orbits situated at fixed distances from one another but not intermediate orbits. Nevertheless, the physicist experiences undeniable satisfaction in attributing a large number of diverse phenomena to this fundamentally unreasonable idea, explaining them by a reality he finds absurd, or at least inexplicable.

Von Laue, the renowned theorist responsible for using the hypothesis of the molecular crystal lattice (based on Bravais's lattice) in research with x-rays — which conception was so brilliantly confirmed by von Laue's own experiments and those of the Braggs — has related in his recent work on the *Theory of Relativity* how he had at first adopted a clearly skeptical attitude toward quantum theory and how later, "the more deeply he had gone into the theory, the more he had felt its inherent persuasiveness."[8] Similarly, Planck maintains that "no one who has studied the lines of these new methods can long resist the spell cast by them" (PRB 60; Eng. 43).

This is indeed the attitude the scientist actually takes in such cases. It does not matter that his reason initially recoils before the postulates a theory would impose upon him. Provided the deductive reasoning is tight enough, he need only follow its development to be won over by the very fact that there is a deduction. Kant was undoubtedly mistaken in his belief that once rational understanding had been achieved it could be definitive and that partial deductions were only links that would inevitably be welded some day into a great *a priori* chain joining all phenomena together everlastingly (cf. ES 2:142 ff., 199). Nevertheless, if one only corrects Kant's ideas with this in mind, one arrives at an exact understanding of the role mathematical deduction plays in science and realizes why the physicist finds it so seductive.

131. *Its Appeal for the Relativist and for the Hegelian*

Here, the mental attitude of the relativistic physicist is indeed closely related to that of the Hegelian. Since it is the former who is our contemporary, however, perhaps the way he behaves will seem more natural to us. We are not too surprised to find that as soon as one precise experiment or another confirms the predictions of a theory, the relativist forgets the arduous effort of the imagination the theory required of him. And this allows us to understand by analogy why a Hegelian, believing he saw how evolution, in the economic realm for example, could be understood as a march toward the realization of an idea, was not too concerned to know whether Hegel had sufficiently justified the transition between thought and physical reality.

132. The Advantage of Scientific Concepts

Obviously, this is not to say that they were equally correct, that is, that the scientist's and the Hegelian's reasons were equally sound. On the contrary, it is quite clear that relativity theory has a great advantage over Hegelianism here. It is the fundamental advantage scientific concepts undeniably have over purely philosophical ideas: they are much more precise. Any theory, any deduction at all, is valid only because of its rigor. A 'flexible' theory, allowing us to account for any observation whatsoever, is useless, even for simple prediction, and is all the more incapable of providing any explanatory power. This is the point of the well-known laboratory dictum: a good theory must be able to be refuted. The quantitative structure of physics — the fact that it measures everything — allows it, and even imposes upon it, a great precision. As a result, experimental confirmation gives physics a power and solidity that mental endeavor outside this structure could hardly hope to attain.

In many European languages it has become customary to use the term *exact* sciences to designate all the mathematical and natural sciences. This expression is not current in French, but common usage provides even more striking evidence of an analogous conviction. They are referred to quite simply as *the sciences* — that is to say, the sciences strictly speaking, as opposed to the science of history, the science of law, the science of sociology, etc., which could not be called sciences except in a broader sense. It is clear, moreover, that what is behind this distinction is the facility with which each of these branches of human knowledge provides verification procedures for its theoretical predictions. It is in this respect that physics is (and, Auguste Comte to the contrary, always will be) more *exact* than sociology.

Of course, even the most carefully constructed scientific theory is never absolutely rigorous. There is always a way to skew it a bit. One has only to study the history of outmoded theories to realize what strange compromises their defenders countenanced before the theories were finally abandoned, what complicated hypotheses they constructed around the principal hypothesis in an effort to shore up the shaky edifice. It is possible that the theory of relativity will not escape this fate. One can already see in Eddington how, in the event that the observation of the solar spectrum did not confirm the predictions formulated by the theory, the relativistic physicist could nevertheless manage to protect the framework of his spatial norms (STG 129 ff.). But if this were actually to occur, one can be sure before the fact that the relativistic conception would lose much of the prestige it now so justifiably enjoys. And if failures of this kind were to recur, even if they could be explained away, it would not be long before the theory was discarded. Undoubtedly then, Einstein's theory offers a considerable advantage over Hegel's in this respect: just as one could more easily be mistaken in the belief that one understood deduction by Hegelian concepts than is possible in the case of the mathematical deduction characteristic of relativism, it was immeasurably easier for the master and his dedicated followers to *cheat* more or less consciously in their demonstrations (if the reader will allow this disrespectful word, which is, in this case, only the expression of an innate human tendency). And it is, moreover, clear that these two circumstances are intimately related.

133. *The Success of Relativism*

This explains why, at the present time, Einstein seems to the initiate to have completely succeeded where Hegel had failed so utterly, namely, in the rational deduction of the physical world (an absolutely essential task, it should not be forgotten, from the point of view of the author of the *Phenomenology* himself). Einstein did not achieve this by following in Hegel's footsteps. He followed Descartes, whose undertaking was, as we know, quite analogous in this respect to that of the German philosopher, and in so doing largely realized Descartes's program of reducing the physical to the spatial. It is clearly impossible to carry out this task completely, and thus the success of relativism is only partial. It is nonetheless real, indeed incomparably more real than that vaunted by Hegelianism; it is also more convincing than that of Cartesianism. Although Cartesianism is absolutely sound in principle (it is in this sense that Descartes remains an eternal model, immeasurably surpassing all those who have followed or will follow in his footsteps), in the detail of its deductions it too often arrived at only very general formulas that made it easy to account for any evidence whatsoever furnished by experience.

It could, of course, be objected that this is perhaps only an illusion due to the spirit of the times in which we are writing, and that if one mentally placed oneself at a corresponding point in the evolution of the Hegelian or Cartesian systems — that is, just a few years after either came to light — one would have a similar impression. We shall return to this question in Chapter 22, where we shall more carefully consider the relationships between Cartesianism and relativism in this respect. It seems to us, however, that the analysis we have just undertaken already allows us to affirm that there are elements in the evolution of relativism that are not ephemeral, since, quite apart from the question of the degree of confirmation the theory can obtain from experimental observations, they afford insights into the relationship of the mind to reality in general. Should the relativistic conception collapse tomorrow (a possibility we considered in §121), it will remain no less true that physical theory at a given moment in the evolution of science believed it could make the physical world largely rational by reducing it to the hyperspatial.

NOTES

[1] Fritz Reiche, *Die Quantentheorie, ihr Ursprung und ihre Entwicklung* (Berlin, 1921), pp. 114, 153–154 [*The Quantum Theory*, trans. H. S. Hatfield and Henry L. Brose (London: Methuen, 1922), pp. 88, 119–120].
[2] Reiche, pp. 114, 115, 117, 119, 124, 125, 138 [Eng. 88, 89, 90, 92, 96, 97, 106]. Langevin rightly considers the way Sommerfeld succeeded in deducing with great exactitude the hyperfine structure of spectral lines, by substituting relativistic mechanics for the Newtonian mechanics used until then by Bohr, to be a "remarkable confirmation" of the theory of relativity (PR 42–43), and Hans Thirring speaks of the "surprising triumph" of this concept (*Die Idee der Relativitätstheorie*, Berlin, 1921, p. 88).
[3] Wien, WW, 122, 167, 233, 294; cf. 148, 298, 300. At the *Conseil de Physique* held in Brussels in 1911, Henri Poincaré had pointed out the disturbing nature of these concepts, which require that one and the same theory be based "sometimes on the principles of the old mechanics and sometimes on the new hypotheses that are their negation," adding that there "is no proposition that cannot be easily demonstrated if only one introduces two contradictory premises into the demonstration" (Paul Langevin and Louis de Broglie, *La théorie du rayonnement et les quanta: Rapports et discussions de la Réunion tenue à Bruxelles, du 30 octobre au 3 novembre 1911 sous les auspices de M. E. Solvay*, Paris, 1912, p. 451). On the internal contradiction involved here, cf. also Edmond Bauer's excellent work on *La théorie de Bohr, la constitution de l'atome et la classification périodique des éléments* (Paris, 1923), pp. 5, 25, 29.
[4] PRB 146, 162, 164; Eng. 100, 112, 113. [The bracketed insertion is Meyerson's. The last quotation, translated here from Meyerson's French, is omitted in the Jones and Williams translation. See p. 146 of the German edition for the original.]
[5] Sir Ernest Rutherford, 'The Electrical Structure of Matter,' *The Times*, 13 Sept. 1923, p. 16, columns 3 and 4 [the bracketed insertion is Meyerson's].
[6] Alfred North Whitehead, *The Principle of Relativity, with Applications to Physical Science* (Cambridge, 1922), p. 3. Cf. also a similar position held by Planck (§ 166 below).
[7] Nernst, although he thought it necessary to attempt to make quantum theory "more accessible to our understanding" or at least to provide some indication of the direction in which we should seek the key to its intelligibility, admits that in the meantime the theory is nothing more than "basically a calculating device, and, one might add, a most peculiar and even grotesque one." Bauer speaks of its "strange consequences" and likewise judges it to be at the present time only "a practical rule which is simply superimposed on the old theories, even though it contradicts them at many points." Fabry

calls it "a peculiar hypothesis" and judges it to be "hard to understand, so strange that it cannot be considered definitive or even the definitive basis for a theory." Lord Rayleigh states that for him it is "hard to consider it as giving an image of reality" (Walter Nernst, 'Sur quelques nouveaux problèmes de la théorie de la chaleur,' *Scientia* 10 (1911) 303; Edmond Bauer, 'Les quantités élémentaires d'énergie et d'action,' *Les idées modernes sur la constitution de la matière* (Paris, 1913), pp. 139, 147; Charles Fabry, 'Les atomes lumineux et leurs mouvements,' *Scientia* 18 (1915) 374, 376; Lord Rayleigh, 'Lettre à M. Nernst,' in Langevin and de Broglie, *La Théorie du rayonnement et les quanta*, p. 50. One will find in this last work, *passim*, an account of the formidable difficulties encountered by Planck's hypothesis). Cf. also what Wien (WW 135) has to say about the particular difficulty in understanding due to the fact that Planck's elements [of radiant energy] must be considered to be in inverse proportion to the wave length of a particular wave. Planck himself, moreover, observes that his quantum of action, unlike atoms and electrons, cannot be "imagined," so to speak (it lacks *Anschaulichkeit*), and he admits that this fact is a considerable disadvantage (PRB 128 [Eng. 87: "precludes ease of representation"]). In a more recent discussion of his hypothesis, the author of quantum theory concludes that to get from the quantum of action to a true quantum theory might take as long as it did to get from the discovery of the speed of light by Olaf Römer to the establishing of Maxwell's theory of light (PRB 163; Eng. 112).

[8] Max von Laue, *Die Relativitätstheorie*, Die Wissenschaft, vol. 68 (Brunswick) 2:7.

CHAPTER 17

THE DEDUCIBLE AND THE REAL

134. *Eddington's and Weyl's Doubts*

In Chapter 11 we noted how surprised and even a bit disturbed the relativist is to witness the disappearance of reality, even though this is the direct consequence of his deductions. This surprise and uneasiness sometimes reach considerable proportions and lead to consequences that are worthy of note because they can enlighten us on the very essence of the concepts we are investigating.

Eddington (as we recounted in Chapter 13, §109) is so struck by the observation that the human mind seems to experience great difficulty in discovering laws "over which it has had no control" (that is, where the mind and reality are seen to be at variance) that he finally asks himself if it is not "possible that laws which have not their origin in the mind may be irrational" to the point that "we can never succeed in formulating them" (STG 200).

This markedly pessimistic attitude cannot fail to surprise us at first, especially if we remember the triumphant declarations (cited in §90) about the universal deduction of phenomena by relativity. His attitude is not atypical, however. Weyl, after stating that the theory allows us to fathom "the nature of the physical world, of matter and of natural forces," suddenly does a kind of about-face, explaining on the following page that

physics does not have a significance with respect to reality beyond that which logic affords in the realm of truth. .... Physics concerns itself only with what, in a strictly analogous sense [analogous to the relation existing between formal logic and truth], could be called the formal constitution of reality. Its laws could never be broken in reality, just as there exist no truths that do not conform to logic; but these laws do not affirm anything about the essential content of reality, they do not go to the heart of this reality.

Indeed, "nothing of the *content* of the reality we immediately perceive enters into the world of physics." Physics and geometry are merged, but "in the last analysis this physical reality appears only as a pure and simple form. ... The entire physical universe has become a form that derives its content from realms quite different from *physis*."[1]

132

135. *The Contradiction*

Let us frankly admit that we are somewhat perplexed on this essential point − actually the most essential point of all, since it concerns the relationship of relativity theory to reality. In our attempt to extract the specifically philosophical content of the doctrine, we have set out to follow, insofar as possible, the expositions of the most authoritative scientists in this field. Thus for the most part we have presented a series of quotations from their writings − a procedure calculated to keep us from going astray − and we thought we could see a perfectly clear and definite concept emerging from them. However, this is no longer the case here. On the contrary, we are up against something that has all the appearances of being a real contradiction. Let us bring it out into the open in order to try, if not to make it disappear, at least to explain it to some extent.

136. *Is Relativistic Reality Reverting to the Self?*

In Chapter 5 we believe we were able to establish that relativity theory, like all explanatory theories in physics, is essentially realistic and consequently postulates the existence of a reality outside our consciousness, different from perceptual reality but nevertheless supposed to explain it. We also allowed that relativism goes further in this direction than earlier theories, that it sets out to strip its *universe* more completely of everything Planck would term "anthropomorphism," that is, everything subjective. Now, by virtue of Weyl's statement cited above − and also to some extent Eddington's statement, if one pushes his affirmation of the subjectivity of physical laws to its logical conclusion − physical reality seems, on the contrary, to be reverting to the self, leaving beyond its reach an obviously still more real, or even the only real, reality − which (at least for Weyl) is no longer of a physical nature. Without trying to fathom the nature of this true reality, about which we are not told a great deal, let us simply remark that we find it difficult to see how it could involve the *qualitative* content of sensation. Such a concept, which would reintroduce quality into reality as Hegel had done, would conflict altogether too much with the powerful general tendency that has dominated science since Democritus, precluding the entry of the qualitative into its domain or (what amounts to the same thing) relegating it to a restricted and impenetrable domain, by attributing it to "the specific energy of the nerve." This concept would turn Planck's pronouncement (§6) completely around, declaring that as soon as we can conceive of measuring something it is not real.

Since reality would reside only in the "immediate date of consciousness," this would obviously tend to draw Weyl closer to Bergson, or even to attribute to him a sort of extreme Bergsonianism, which would perhaps not be to either Bergson's or Weyl's liking.

However, even in the case of physical reality we are still greatly perplexed, for we must choose between two points of view that seem to be impossible to reconcile.

137. *The Subjectivistic Affirmations are Beside the Point*

Let us hasten to add that the choice seems obvious to us. It is impossible to read through an exposition of relativistic physics without noticing that the universe whose existence is postulated is indeed conceived to be real; every page, every line is eloquent testimony to this, and the demonstrations would lose all meaning if we tried to make the existence of the universe depend on the self for even one moment. This is the case even in the two books by Weyl and Eddington, as the reader has come to see by the many citations we have taken from them. The subjectivistic affirmations, on the contrary, only come at the very end and are quite brief; it is easy to see that they in no way follow from the main body of the reasoning, that they constitute a sort of aside. By removing them one would in no way damage the structure of the work. One might almost say that they appear rather as simple flights of fancy.

138. *Where Does this State of Mind Come From?*

Furthermore, the state of mind they express is not too hard to understand. We have indeed seen above that the relativists, unlike Descartes or Hegel, do not undertake universal deduction with any preconceived goal in mind. Although they certainly carried out this undertaking with admirable logic and in accordance with rigorously observed rules, the final goal seems to be so vaguely perceived from the beginning that the fact of having achieved it comes as rather a surprise to the very theorists who contributed so much to the success of the work. And they are also apt to feel the same surprise in considering the very result they had worked so hard to attain. They are in some sense disappointed. Was it really as simple as that? Did all this bewildering multiplicity of phenomena result from nothing more than the combination of a small number of spatial elements? Were Maxwell's equations — established with so much effort by the great scientist over the concepts of Laplace and Coulomb that dominated the physics of electricity at that time – merely the simplest invariant form of the relation between four parameters, and could they therefore have been formulated prior to any experimental work by a simple *a priori* deduction?

We saw above (§116) that this suggestion is only partially correct and that in reality pure deduction is likely to produce only the mathematical framework of these equations and not the propositions themselves. But we also know that the relativist is inclined to forget this very fundamental distinction, and it is in fact his psychology that concerns us here. Descartes was able to believe that his elementary matter – in spite of the fact that it was nothing but space – was sufficient to account for all the phenomena of the universe, and, likewise, Hegel was able to claim boldly that he had reconstructed reality by means of his concepts; this is because they genuinely wanted to accomplish this impossible feat. But the relativist, on the contrary, is somehow inclined to recoil when he contemplates his own work.

This is a state of mind with which we are already acquainted, since we considered a manifestation of it in Chapter 11, namely, Eddington's surprise at the disappearance of matter.

139. *Weyl and Schelling*

Did Descartes or Hegel sometimes harbor reservations like those manifested by Weyl and Eddington? If they did, they certainly did not take us into their confidence. In their public writings both of them were invariably and unshakably dogmatic from the beginning to the end of their careers. As a result, if we really want to find a parallel to the surprise that troubled Weyl when he saw himself transported into an empyrean from which reality appears only hazily in the distance, we must, as we have pointed out (§91), look not to Hegel himself but to a disciple who, despite his fervor, nevertheless cannot protect himself from all doubt.[2] Similarly, one finds a parallel to Weyl's statements concerning the purely formal and logical import of relativistic physics, not in Hegel but in Schelling – Schelling in his second period, of course, when he takes a stand against his disciple in an attempt, if not to demolish his work, at least to diminish its significance. He argues that the immense Hegelian synthesis is in reality nothing more than a pure and simple logic, and that all the philosopher's attempts to go beyond these limits are doomed to fail deplorably:

As could be predicted, the logical motion inherent in the notion sustains itself within the limit of the purely logical; but as soon as it must leave this realm to move into the domain of reality, the dialectical movement stops and breaks down.

For "all these so-called *a priori* forms express only the negative side of all knowledge — that without which no knowledge is possible — and not its positive side — that by which knowledge comes about. . . . Consequently, their universal and necessary character is only a negative character," that is, a character indicating "that without which nothing is, but not that by which something exists." What Hegel should have done, Schelling says on another occasion, when he feels in a bit more indulgent mood toward his former friend, was to make logic, not just part of his philosophy, but his entire philosophy; it ought to have been "absolute logic, the ideal of the pure science of reason."[3]

140. *The Opposition Between Thought and Reality*

What we see here is something deep and far-reaching; indeed it is none other than the eternal opposition between thought and reality. In science, as we noted in Chapter 11 (§100), this conflict manifests itself by the coexistence of a tendency toward idealism side by side with realistic convictions. But this contradiction, of course, is not peculiar to physical science; it precedes physical science and originates in the innermost recesses of our minds. We would certainly *like* reality to be rational, but at the same time we *feel* that this desire is essentially chimerical, that it cannot be completely satisfied. We feel this so strongly that, although we make an effort to discover the rational in the real — to demonstrate that reality is entirely rational or, inversely, that reason alone is capable of constituting reality — as soon as we believe we have more or less attained this goal, any part of reality we have shown to be rational is apt to seem to us not to belong to true reality. There is an element of paradox in our minds here; in order to establish its existence more clearly, we shall take the liberty of illustrating our claim with two examples taken from areas far-removed from the physical sciences.

141. *The Characters of the Novelist and the Playwright*

When a novelist describes a character, his primary aim is certainly to make us understand the psychology of this character, that is, to show us how his actions follow intelligibly and logically from a certain number of basic character traits that the author either describes explictly or suggests. At the same time, a good writer knows that if he made his character too logical or intelligible, he would succeed only in creating a marionette, something mechanical, devoid of any semblance of life. The semblance of life can, on the contrary, be breathed into it by adding to what might be called the logical concept or general outline of the character a certain number of little traits that are independent of this framework, that is to say, extralogical. The author will use his talent to blend these disparate elements, to make us accept all of them taken together as representing the indivisible whole that is the character and spirit of a man, an *individual*. Balzac, Maupassant, the Russian short story writers (and especially their ancestor Gogol) excel in this difficult art. By analyzing one of the concise short stories of Maupassant, for example, we can see the surprising force with which he succeeds in establishing and animating the most unlikely character, using only the smallest number of apparently unimportant details. But the incontestable master is Shakespeare; he manages in this way to make real men and women out of characters who would seem to be set apart from all humanity by the extravagant tragedy of their fate. He was so successful that perhaps his most lifelike character, Hamlet, is certainly not entirely intelligible. We do not always see clearly the motives behind his actions; he appears somewhat opaque, like someone we might have known in real life. In short, in examples such as these, what we conceive as being the innermost essence of reality — which coincides here with the individual — actually seems to reside in the nondeductive, the nonrational.

142. *The Historicity of Jesus*

We shall take our second example from the religious sciences. It is well-known what importance Christian apologetics has, from the time of the earliest church fathers, attached to demonstration founded on the predictions of the Jewish prophets. Through the immense labor of countless theologians, all the circumstances of the life of Jesus, even the seemingly most unimportant, were thoroughly explored, and it was carefully established that each of them corresponded to one Old Testament verse or another, in short, that it constituted only the translation into fact of the verse in which it was *prefigured*. We know, too, that as late as Pascal, for example, this demonstration seemed to constitute by far the most powerful argument that could be invoked in favor of the truth of the Christian religion.

The situation changed entirely as soon as audacious critics attacked the historicity of the very personage of Jesus, declaring that the image furnished by the Gospels had in fact been made out of whole cloth in order to make him appear to be the Messiah awaited by the Jews. And even supposing that the man had existed, they added, he was a person about whom nothing could be known, "a man without a biography," since all that was reported about him referred to his Messianic activity. Thus, by a reversal not unlike others to be found in the history of human thought, the whole immense body of scholarship that Christian apologetics had acquired over so many centuries now served to provide arms for its adversaries.

What was the attitude of the defenders of the tradition under these circumstances? Obliged to meet an attack coming from a direction diametrically opposed to the one from which they were usually attacked — since until then critics had merely contested the divinity of Jesus while fully admitting his *human* existence — they too took aim in the opposite direction, that is, they attempted to show that not all the details of Jesus' life could be deduced from the prophecies. They admitted that some circumstances were even quite inconsistent with the orthodox idea of the Messiah; that they seemed to be genuine anomalies with regard to this Messianic role. Schmiedel, who reduced this line of argument to a system, counts nine principal circumstances of this kind, and it is on these "nine columns" that this apologist constructs his demonstration of the historicity of Jesus.[4]

Thus, even more clearly than in the case of the men of letters discussed above, the reality of the character is tied here to characteristics that do not lend themselves to *a priori* deduction; it depends on *extralogical* — singular — traits. This opposition between rationality and existence is already found in Saint Thomas (who was, moreover, only commenting on an Aristotelian concept): *existentia est singularium, scientia est de universalibus.*[5]

The reader will now understand why Weyl hesitated to reduce all physics to space, why he believes that what has thus become entirely deducible cannot be the true reality.

143. *This has Nothing to do with the Unpredictability of the Vital*

It could, however, be objected that our two examples are not very convincing, because they concern phenomena having to do with life, and even human life; therefore, all they reveal is the fact that we have a deep-seated conviction that manifestations of life (and more particularly human life) cannot be foreseen, that they involve free will. From such a point of view, what is involved in these examples is prediction or lawfulness, not comprehension or rationality as in explanatory physical theories.

Let us begin by observing that it seems, at the least, risky to claim that our understanding instinctively and immediately makes a fundamental distinction between vital phenomena and all other phenomena when it comes to applying scientific norms. In everyday life, we obviously attempt to predict and explain both of them indiscriminately, and these constant efforts seem to be evidence that we consider both of them predictable and explicable on more or less the same grounds. The same is true most of the time in the domain of conscious thought. Not only does so-called naive thought not make any distinction between men and animals insofar as free will is concerned, but it generally assimilates the entire physical world with the world of living things, and this *hylozoism* has, as we know, appeared many times in diverse forms in the history of ideas. On the other hand, from the beginning of philosophy and science, theories have been put forward postulating the absolute necessity and complete rationality of all phenomena. What seems to depend on human volition is explicitly included in this affirmation. It would, therefore, be quite strange if, in the case of our two examples, understanding unconsciously and unhesitatingly applied a distinction that otherwise seems so uncharacteristic.

Moreover, the perfect analogy we observe here between such different examples is, in itself, of a nature to make us suspect that they are all manifestations of the same thing. Consequently, what the relativists' attitude shows us is that it is not the assumption that the phenomenon is predictable that shocks our deep-seated feeling, but the assumption that it can be completely reduced to rational elements, that it is constituted only of these elements.

144. *There are Degrees of Rationality*

It will perhaps be recalled that we already ran into this opposition between the rational and the real in Chapter 3. There (§28) we reached the conclusion that even the geometrical is not purely rational in nature. On the contrary, if contains what must be considered a qualitative element, and it is undoubtedly due to this circumstance that we are able to use it to explain reality. Thus, the geometrical passed for reality to a certain extent, while here it appears as something rational. But this is because there are degrees of rationality (and consequently of reality as well, if we consider the two terms to be opposites); if geometry is less rational and more real than the purely algebraic, it is more rational and less real than the physical. It is just real enough to be able to deceive the mind to a certain degree and consequently to serve as a representation of reality, but not real enough to prevent us from having misgivings when we try to substitute it completely for reality. All these reasons taken together certainly tend to convince us that, in the final analysis, it is indeed the nondeducible — that which, in the words of Kant quoted above, must be given but cannot be understood — that seems to constitute the essence of reality.

145. *The Conflict between Realism and Acosmism*

Furthermore, a restatement of this observation, insofar as it relates to the contradiction observed in the heart of the relativistic doctrine, might make the deep and almost inescapable implications of these examples stand out a bit more clearly for the philosopher. Science is realistic; we know, however, that if explanation is pushed to its limit it can only end up at acosmism — the destruction of reality. In relativism, precisely because it is a very advanced, very perfect form of theoretical explanation, these two extremes of existence and nonexistence are found to be quite close to each other. This leads to a painful sort of conflict for the physicist.

146. *Nonrational Laws*

Eddington is obviously undergoing this kind of conflict when he supposes that laws of reality not originating in our mind might be entirely unintelligible to us. For there is no apparent reason why the mind cannot conceive a law as a pure and simple rule describing how things happen and designed to predict how they will happen in the future, even though this rule can in no way be deduced from the principles governing our mind. Surely in everyday life we constantly formulate rules of this sort: we do not understand why things happen in such and such a way, but we *know* that they do and that we must act accordingly. As for science, although positivism is wrong to suppose that it is concerned only with effects and consequently presents only a collection of rules deduced from experience, it is certain that it very often begins with such rules and only later attempts to deduce them rationally. Thus Maxwell's equations, which summarize the body of laws obeyed by electrical phenomena, were surely nothing more than purely experimental principles before the advent of the theory of relativity. They were rational to such a slight degree that they actually contradicted what then appeared to be the only possible basis for a deduction, namely, mechanics.

It is true, however, that this might be only a deceptive appearance. Maxwell's equations seemed impossible to deduce, but they were really only the simplest invariant form of the relation of four parameters (§138). Thus, even though neither Maxwell nor those who followed him were the least bit aware of it, the rationality was explicitly there, and it is only because of this fact that these laws could be formulated. From a more general point of view, it must be pointed out that, although the scientist seems to be seeking a simple empirical rule, in reality he cannot do so without being guided by a theoretical vision, that is, an idea in which the element of rationality necessarily plays a part (cf. ES 2:289 ff.). It is this rational element that makes it possible for the rule to be formulated. The rule doubtless appears to us to be empirical, but it is in fact rational, and the future development of science will show this to be true.

Explained in this way it is obvious where such a thesis is headed: it is an affirmation of the complete rationality, if not of nature, at least of all science, and is thus the exact opposite of the positivistic doctrine. It is doubtless correct to say with Goblot that "we cannot be satisfied with truths of fact, we need truths of law."[6] But this is, after all, only a postulate, a *desideratum* of our minds: we wish it were so. To maintain, however, that it is really so — to declare with Plato that where there is no rationality in nature there can be no regularity — appears very risky, to say the least, in the present state of science. For we finally know some of these laws that seem fundamentally incapable of being reduced to the canon of reason. Carnot's principle is one of them: to deny this fact is to assume that one will someday be able to make this whole principle rational, to explain how the universe constantly advances in a single direction without having this evolution originate from an essentially improbable state — and that certainly appears to be contradictory. Atomicity and the existence of quanta or, in more general terms, spatial discontinuity, are similar cases, for we shall never be able to deduce anything that is not continuous.

The truth is that behind this opinion lies a confusion between the concepts of lawfulness and rationality (or causality). This way of thinking is not uncommon among philosophers as well as among scientists, but this does not make it any easier to justify; on the contrary, if one would understand the structure of science at all, nothing is more useful than maintaining the distinction between these two concepts. In the present case, it will be clearly seen that we are dealing with a *given* and must resign ourselves to accepting it as such. We are incapable of deducing it from our reason alone; it is thus imposed on our reason from the outside. This does not keep us from being able to describe it or determine it numerically; therefore it can serve as a basis for the deductions by which we strive to explain reality.

147. *Reality does not Disappear*

Will we be accused of philosophical recklessness if we dare claim that relativistic physicists would do well to bear these elementary observations constantly in mind? Perhaps they would then be less inclined to overstimate their achievement, as considerable as it is; on the other hand, they would surely also be a little less inclined to experience the sort of disappointment at the disappearance of reality that we have seen in Weyl and Eddington. For reality does not disappear; it does not dissolve into mathematical formalism, or at least only slightly. A significant part of it remains — enough, in any case, to guarantee that reality remains real and cannot be merged with our thought. This remaining part takes refuge, on the one hand, in diversity in its two principal forms — namely, change in time, as seen in Carnot's principle, and differentiation in space, as seen in the discontinuity of atoms and quanta — and, on the other hand, in interpretation, the only thing that transforms what is deduced by mathematics into a physical result.

NOTES

[1] Our quotations are translated from the third edition of Weyl's book (*Raum, Zeit, Materie*, pp. 262, 263). This discussion was almost completely suppressed in the fourth edition (the one used by the translators). It goes without saying that this does not diminish the importance of the epistemological conclusions we draw from these statements (conclusions that also follow from Eddington's quite similar views), but it does perhaps show that the author himself realized that the opinion he had expressed was not completely consistent with his general position on this point. [We have translated directly from Meyerson's text here; the bracketed insert is his. The first short phrase can be found on p. 311 of the Brose translation (STM); the rest, as Meyerson points out, does not appear in the fourth edition.]

[2] To be fair to Taine, we must note that he himself clearly acknowledged this doubt, and quite wittily so. In fact, after saying that "one would think one had arrived at the summit of human knowledge and the ultimate perspective on the world," he adds: "if one did not see over there on the corner of the table a volume of Voltaire sitting on top of a volume of Condillac."

[3] Schelling, *Jugement sur la philosophie de M. Cousin*, trans. J. Willm (Paris, 1835), pp. 17–19; Kuno Fischer, *Geschichte der neueren Philosophie* (Heidelberg, 1899), 7:266.

[4] Paul Wilhelm Schmiedel, *Die Person Jesu im Streite der Meinungen der Gegenwart* (Leipzig, 1906).

[5] Cf. Zeferino González, *Histoire de la philosophie médiévale* (Paris, 1890), 2:254 and Aristotle, *On the Soul*, Bk. 2, Ch. 5, 417^{b}23: "what actual sensation apprehends is individuals, while what knowledge apprehends is universals" (trans. J. A. Smith).

[6] Edmond Goblot, 'Sur le syllogisme de la première figure,' *Revue de métaphysique et de morale* 17 (1909) 357–366.

CHAPTER 18

THE SYSTEM

148. *The Relativists and Kant*

The relativistic deduction has pushed the scientific effort so far, it has so clearly opened up what constitutes the common ground of science and philosophy — namely, the question of the ultimate nature of reality — that almost since its appearance, and in any case well before Einstein gave it the form of *general* relativity that we have examined here, attempts have been made to fit it into a specific metaphysical position. As the theory has developed, these attempts have become more numerous and have taken on more and more importance, to the point that physicists themselves have sometimes joined in, declaring themselves in favor of this or that *system*. Although these opinions certainly deserve to be taken quite seriously by the philosopher, it goes without saying that he cannot give them the same weight he gives their opinions concerning the strictly scientific side of the theory. However qualified their authors may be in the domain of relativism, these opinions are only testimony or documentation as far as the philosopher is concerned. He must carefully take note of them, but he must examine them in the light of his own methods.

Kant is certainly the name mentioned most often by the relativists themselves. As a matter of fact, those who have concerned themselves with the philosophical aspects of relativity have referred to him almost exclusively. Einstein himself, however, seems to be an exception. As far as we have been able to determine, he has made no explicit declaration of this sort, which could perhaps be explained by the quite natural reserve of a scientist who hesitates to go beyond the limits of his own field. Furthermore, at a meeting of the Société Française de Philosophie we heard him express quite explicit reservations in response to Léon Brunschvicg, when the latter invoked the name of the Königsberg philosopher in connection with relativism.[1]

These reservations seem to us to be justified, despite the fact that other experts in this field have expressed the same opinion as Brunschvicg. Cassirer in particular made this position the basis for his altogether remarkable book on relativity theory, which we have often cited. This position is clearly far from prevalent among philosophers, however. Below, for example, we shall see Reichenbach's very radical opinions on this subject, and it would seem obvious that most philosophers, at least in Germany, are more inclined to see a conflict between the ideas of the master of criticism and the relativistic concepts; they attack relativity in the name of Kantian *intuition*. [2] This is an opinion to which we shall return shortly.

Those who compare Kant and relativism most often stress the concept of time. Relativism, they tell us, shows that we cannot assume that the notion of time as we find it formed within ourselves can be the property of things themselves; it is thus only the form our mind attributes to them, which is certainly consistent with Kant's position. And they go on from there to state that, in general, the concept of reality resulting from relativity theory agrees completely with what Kant called *transcendental idealism*.

149. *Kantian Space and Time as Forms of the Mind*

Beginning with time, let us say that the analogy they make, though real, is in fact superficial: we would go so far as to say that the principal reason so many fine minds have been seduced by this position is the strictly extraneous circumstance that the doctrine was first presented in the form of *special* relativity (§42). Indeed, if one examines the question a little more carefully, the differences leap out at us. It would, of course, be impossible to give too much credit to Kant, who by the sole power of his thought, that is by reasoning abstractly and not — like the relativists and those who follow them — from concrete observations, was able to raise himself to such heights that what seems most essential to us in the reality around us, namely, its framework as constituted by space and time, appeared to him as simple forms imposed on things by the mind. This was all the more impressive since, as we know, he derived all his scientific concepts from Newton, for whom the existence of absolute space and time was a sacrosanct dogma. One might say that, by thus breaking the rigid framework to which reality was confined by the Newtonian concept, the Kantian critique to some extent prepared our minds for the coming evolution. And it is surely quite remarkable that Kant's speculations in this area came two generations before those of Gauss, Lobachevsky and Bolyai, in which the possibility of a configuration of space different from that of naive perception was envisaged for the first time. It must be admitted, however, that these geometricians were in no way influenced by the Kantian ideas and that their true point of departure was altogether different.

150. *Relativistic Time and Space as Separate from the Self*

Turning now to a comparison of Kant's concept with that of relativism, let us point out that, while critical philosophy begins by separating time and space from the object, from the thing-in-itself, and attaches them to the subject, relativism is the expression of the opposite tendency. From the beginning, relativism sets out to give everything having to do with determinations of time and space a form equally valid for all observers. This clearly can be done only by separating them from the subject and giving them an existence independent of the self, that is to say, by putting them back into the thing-in-itself. Thus, time as we ordinarily understand it is indeed the property of our self. But what remains of time after we have transformed it into a fourth dimension of the spatio-temporal universe – and what no doubt constitutes its most essential part according to the theory – belongs, on the contrary, to this four-dimensional universe, understood to be independent of the self and situated outside the self. And the same is true for space: Euclidean space belongs to us in our own right, but relativistic space filled with *curvatures* or *puckers* is objective; it belongs to things, or rather it is the very essence of the objective. It alone constitutes all things; it is (in spite of the 'flights of fancy' we pointed to in the preceding chapter) truly and literally *the* thing-in-itself.

We do not believe the attentive reader can be in any doubt that this is really the spirit that inspires the doctrine. Of the many passages we have cited to this effect he need only recall the one in which Weyl declares that the problem of physics is "to get an insight into the nature of space, time, and matter so far as they participate in the structure of the external world" (§47).

Cassirer, in an attempt to show that relativism conforms perfectly to Kant's system, insists quite naturally on the fact that for Kant, space, according to the well-known phrase, "is not an object." It obviously is an object for the relativists, as follows if one but considers that the doctrine – as we have amply demonstrated and as Cassirer himself also admits – tends toward an *objectification* of reality,[3] while the only objective factor at which it arrives is manifestly spatial. This is not related in any fundamental way to the Kantian concept, and it seems to us that we only tend to obscure the real thrust of relativism by invoking the master of criticism in this case. With all due respect to the writings of such distinguished scientists, we might suggest that some clearly nonscientific statements on the part of these leading physicists must have been inspired to a certain extent by the desire to associate themselves with a philosophical point of view that is in fact quite foreign to the relativistic doctrine.

151. *The 'Copernican' About-Face in Kant and in Relativism*

It must be admitted that Kant himself sought to justify the objectivity of mathematics at the same time that he attributed time and space to the self. For him this objectivity follows from the fact that, since objects can only affect our sensibility by submitting themselves to laws conditioned by the forms of this sensibility, it is therefore these laws — which are the laws of mathematics — that constitute the framework to which reality must invariably conform. And undoubtedly this concept can to some extent be adapted to relativity theory by a sort of superposition. Once relativistic reality has been separated from the self, it is not beyond the resources of philosophical dialectics to perform the famous critical or 'Copernican' about-face by making this reality, too, revolve around thought. This is indeed what Cassirer has attempted to do, and Weyl seems to have followed a similar train of thought in stating, as we have seen, that physics is concerned only with "the formal constitution of reality." One must be aware, however, that the attempt is carried out under conditions quite different from those envisaged by Kant; it is much more complicated and consequently more difficult. For Kant himself returned everything dealing with the temporal and the spatial to the subject in a single step. In Kantian relativism, on the contrary, there would be two quite distinct stages: in the first stage the time and space of our naive intuition would be declared subjective, with the four-dimensional universe of relativity remaining independent of the self; only in the second stage would this universe in its turn be reintegrated into the self. Furthermore, to carry out the second change one would no longer have at one's disposal a foundation as solid as the one furnished to Kant by the temporal and spatial intuition just mentioned. One would have shaken this foundation oneself by the criticism constituting the first stage.

152. *Temporal and Spatial Intuition*

Intuition itself, moreover, given the task of imposing the temporal and spatial framework upon reality, would become in some sense a two-stage mechanism. There would first be the intuition that immediately perceives Euclidean space and ordinary time. Behind it there would be another intuition, purely mathematical in nature, imagining the four-dimensional universe, to which, in turn, it makes reality conform. It bears repeating that all this does not seem inconceivable, but it does appear rather complex and difficult if one reflects upon it.

If, on the other hand, one considers things with an open mind, one would seem to be led to the position of those who believe that relativity theory tends to destroy the concept of Kantian intuition. Indeed, though Kant recognized that the notion of space furnished by our visual sensation is different from that constituting the framework for our tactile sensations, he affirmed that behind both of these "empirical" conceptions there was one "pure" *a priori* intuition. For the relativist, on the contrary, the spatial conception joining together all the diverse perceptions of our senses into one whole constitutes a simple mental construct, formed with the aid of these very perceptions and intended to establish an agreement between them just as it does between the perceptions of different observers, in the manner explained in Chapter 5 (§49).

This agreement can ordinarily be brought about within the framework of Euclidean geometry, which is the simplest geometry and one that can be maintained as long as we are dealing with summary observations carried out directly with our sense organs. If, however, we refine these organs by the use of more and more delicate instruments, there comes a time when we see that this framework must be modified. But space remaining what it is, that is, a construct, a framework created by our reason, there is no *pure intuition* to prevent us from proceeding with this change.

153. *The Superposition of the Two Points of View*

Up to this time, the significance of these difficulties does not seem to have been fully appreciated by those who would unite relativism with the Kantian concept of space. We might even say that they have sought instead to obscure the difficulties by confounding the two stages we have shown to be so different in nature and by suggesting that relativism is close to the Kantian position by the very fact that it calls into question the certainty with which our perception affirms the independent existence of time and space. Although these two points of view can certainly be reconciled by superposing one on the other, this cannot be accomplished without doing violence to one or the other. Thus, without going quite as far as Reichenbach, who maintains that "Kant's pure intuition *cannot* be reconciled with relativity's doctrine of space and time,"[4] we believe that the least one can say is that the nature of relativism in no way forces a Kantian interpretation.

154. *Mathematics in Relativism and in Kant*

It should be added that, by assimilating relativism into the Kantian position, one seems in some sense to be abandoning the most marked — and we might say the most valuable — characteristic of the modern theory, namely, the exclusively mathematical nature of its deduction. Indeed, mathematical deduction is, as we know, what characterizes science, and we pointed out at the beginning of this work (§1) that all attempts to turn it from this path have ended in failure. From this point of view, relativism is obviously a sort of *summum*; it is entirely mathematical and nothing but mathematical. With Kant the case is entirely different. He was undoubtedly a 'Newtonian philosopher'. But when it came to actually deducing what he thought could be known *a priori* in the physical sciences — what amounts to the outline or framework that was to form the *Metaphysical Foundations of Natural Science* — rather than proceeding by the method that undertakes "the construction of concepts by means of the presentation of the object in *a priori* intuition," which "is called mathematics," he often preferred the method using "pure rational cognition from mere concepts," which belongs to "pure philosophy or metaphysics" (cf. §97). Indeed, we find very little mathematics in the above-mentioned work. This fact becomes all the more significant when one notes that it exercised a considerable influence on the whole subsequent development of German philosophy. The thinkers of the romantic school — Schelling and his disciples, Hegel — are determined to set up, in the face of the science of the scientists, something absolutely different and even antagonistic: the "philosophy of nature." The most characteristic trait of this construct is the abandonment of mathematical deduction, and indeed even disdain for anything mathematical. This is not the first time in the history of human thought that one has, without questioning the principle of deduction itself, abruptly modified the way in which it was accomplished, by substituting extramathematical deduction for mathematical deduction or vice versa. The most striking example is that of Aristotle, who replaced the panmathematicism of his master Plato by a panlogism (cf. ES 2:141–142). But each time a change of this kind has occurred, the new doctrine, while deriving in part from the old one, has nevertheless openly opposed it. This is not the way the 'philosophers of nature' and Hegel react to Kant, and especially to his epistemology. On the contrary, they explicitly appeal to both Kant and his epistemology; they claim to be only accomplishing what the master of criticism had sketched out. Of course, they are wrong, because mathematical deduction does play a considerable role for Kant; compared to his successors, he certainly seems to be a partisan of scientific mathematicism. However, their attitude reveals that he was, after all, much less imbued with this principle than has sometimes been claimed, and probably less than he himself believed.

155. *Relativism is a Scientific Mathematicism*

But if relativism is not transcendental or critical idealism, if, as we saw at the beginning of this work, it is even less a positivism, what is it from the philosophical point of view? Can it be assimilated into any particular conception of reality, to any metaphysical *system* having a name, if we may thus express ourself? We think such an assimilation is possible. To accomplish it one need only recall the most essential traits of relativism, always keeping in mind that one is dealing with a strictly scientific construct, which as such must differ considerably in some particulars from those conceived by philosophers themselves.

From the very beginning of this book we have stressed many times the fact that relativism is a mathematical theory and that it reduces all reality to mathematics. It is thus essentially a *mathematicism*. If one wishes to make it into a philosophical system, one will have to consider it to be a *panmathematicism* or a *metamathematics*, to use the excellent term Brunschvicg coined to characterize Plato's theories, a term that seems to apply in a particularly felicitous manner to all philosophies of this sort.[5] The only reason relativism was not immediately recognized as such seems to be that it issued from physical science and limited its ambitions strictly to the confines of physical science, leaving aside all the other realms of the mind.

Relativism includes neither ethics nor esthetics. Before attempting to provide it with either, one would first have to give the ethics or esthetics a mathematical form, as Plato tried to do; certainly no relativist has yet undertaken this. If one tries, on the other hand, to accomplish this end by mixing the purely physical theory with a more general system already worked out – and, it goes without saying, lacking none of the parts we expect to see evolve naturally out of philosophical conceptions – one distorts its fundamental meaning.

156. *It is not Empirical*

Furthermore, having grasped this essential nature of relativism, one will be less surprised at the way it has been claimed by the most divergent positions:

not only has it been represented as positivism or critical realism, attempts have also been made to classify it as 'radical empiricism' (cf. Cassirer, ER 94; Eng. 426). The last designation is as unsuitable, perhaps even more unsuitable, than the preceding ones, since relativism is, above all, an essentially deductive conception; consequently (as in everything truly mathematical) the role of experience, of the strictly *empirical*, is greatly reduced in favor of reasoning. But this is because in the field of mathematics the agreement between reason and perception, which we have stressed above (§§23, 97), causes idea and reality in some sense to blend together.

157. *Philosophical Panmathematicism*

But precisely because relativism is a very advanced, and at the same time very pure, form of panmathematicism, it becomes harder to misinterpret it. This can be seen by contrasting it with the properly philosophical theory of Cassirer, who, as we know, is one of the most authoritative representatives of the Marburg school.

In an earlier work we had also described this philosophy as panmathematicism, a designation it no doubt largely deserves. It is, however, a thoroughly philosophical panmathematicism, and as such has embraced, so to speak, a critical idealism borrowed from Kant, or rather developed from Kantian ideas. Far be it from us to criticize here the vast and beautiful structure that has thus been built, and we cannot even dream of doing justice to the rich content of this thought. We shall only point out that, as might be expected from what we have just said, the union of essentially mathematical and properly philosophical elements seems to be a source of considerable weakness, much like the union of thought and extension in Cartesianism. Relativism avoids this serious drawback since, by definition, it precludes everything that is not purely physical.

158. *Geometrism and Algebrism*

Moreover, the panmathematicism of the relativists is basically quite different from that preached by Cassirer, even from the point of view of the underlying mathematics. The former is in fact a pangeometrism, as we have pointed out many times, while the latter would be a panalgebrism. It is true that the relativists sometimes express themselves a bit ambiguously on this point (as we were obliged to concede in §§54 and 120). Weyl, for instance, states that "colours are thus 'really' not even aether-vibrations, but merely a series of

values of mathematical functions in which occur four independent parameters" (STM 3–4). It is only natural that Cassirer should take advantage of
this ambiguity to try to interpret relativism in the light of his own theory. He
declares that the theory of relativity "in a general epistemological regard . . .
is characterized by the fact that in it, more clearly and more consciously
than ever before, the advance is made from the copy theory of knowledge to
the functional theory." In this connection he cites a statement by Kneser, to
which he fully subscribes and according to which the special theory of
relativity has already "substituted mathematical constructions for the apparently most tangible reality and resolved the latter into the former."[6]
Although it is, of course, quite correct to say that the special theory of
relativity partially transforms physics into mathematics, and that general
relativity does so much more completely, this is true only provided that the
particular form of mathematics involved is understood to be geometry. After
all, what is put in place of the physical world must be somehow capable of
being conceived as something real; one must be able to imagine it, at least to
a point. Confronted with pure algebraic functions, on the other hand, our
imagination would stop short (§28).

Cassirer, who himself admits that relativism pushes back "anthropomorphism" (§48), also explains that the evolution toward greater and greater
relativity seen in the theory "represents no contrast to the general task of
objectification, but rather signifies one step in it." While maintaining that "all
its knowledge of objects *can* consist in nothing save knowledge of objective
relations," he concedes that there are "relations, some of which are independent and permanent and by which a certain object is given us."[7] But the
truth is that, as we realized in Chapter 3, all science obviously requires the
presupposition that objects exist independently of the self, and we have seen
that relativism is no exception.

159. *Geometrism as an Intermediate Step*

Thus, from this point of view, and insofar as it is a geometrism, relativism is
a less radical mathematicism than Cassirer's. This helps us to understand
why, as we have noted (§120), we sometimes find statements bordering on
panalgebrism in the writings of the relativists. It is because our tendency
toward deduction is in fact as absolute as it is imperious: we would really like
to be able to deduce *everything* with our reason alone. In the geometrical we
cannot avoid the feeling that there is something given, and for this very
reason our deduction is tempted to go further, to go all the way to the purely

algebraic. This tendency is seen very clearly in Weyl's declaration, in connection with a fundamental coefficient required by electron theory, that "we
regard with . . . scepticism the belief that the structure of the world is founded
on certain pure figures of accidental numerical value" (STM 262). Thus
anything *given* appears to be an accident spoiling the absolute deduction, and
it is no longer surprising that geometrism seems to be only a kind of step
along the way. It is in fact at just this stage that the theory of relativity stops,
even taking into account Weyl's and Eddington's additions. Weyl strikingly
sums up his position on the nature of the relativistic theory of gravitation
when he states that in Einstein's theory we find a sort of amalgam of the
theories of Newton and Pythagoras (STM 228). This is quite apt, because
these physicists are indeed inspired by Pythagorian panmathematicism. Since,
however, Weyl is stressing here the geometrical nature of this panmathematicism, it would perhaps have been appropriate to add Plato's name to that of
Pythagoras, for, as we know, it was he who gave Pythagorianism a geometrical
form. In this context one could also recall (as Weyl himself did, moreover, in
the passage we have cited in §87) that even before Newton, perhaps the most
powerful mind of which humanity can boast — we mean Descartes — had,
with the quite imperfect means at his disposal, undertaken the overwhelming
task of mathematicizing the physical world. It is true he only partially accomplished this — complete success being clearly impossible — but he was successful enough to convince his contemporaries and the next generation.

160. *Plato*

We believe, however, that one can best understand the true nature of the
problem that concerned us in the preceding chapter by going back to the
language of Plato. Since relativism has set out to make physical reality rational, and seems to have largely succeeded, must we assert that everything
in reality that is thus demonstrated to be in accord with our reason cannot be
truly real but must belong to reason alone, which projected it outside itself,
so to speak? This is obviously a purely metaphysical question and as such
can be answered in various ways. But the most natural attitude for the
philosopher seems to be to follow Plato in simply noting the agreement
between thought and reality in mathematics. Relativism, we shall say, is a
mathematicism, and what is mathematical, since it belongs (§97) *at one and
the same time* to our reason and to nature, is neither altogether the *Same* nor
altogether the *Other*, or rather is both at once, being the true *intermediate
substance* (cf. ES 1:28; 2:315). Thus we see in relativism the primordial fact

that, from the standpoint of nature, man is not "a kingdom within a kingdom." On the contrary, "it is impossible, that man should not be a part of Nature," as Spinoza said.[8] Again then, nothing forces us to suppose that what relativism has deduced thereby ceases to be real. While we do not deny that it is possible to superpose on relativism a metaphysics similar to the Kantian theories, it seems to us that if one wants to give relativism an idealistic interpretation, this can be done in a much less artificial way by putting it into the framework imagined by the master of Athens.

NOTES

[1] Cf. meeting of 6 April 1922, *Bulletin de la Société française de philosophie* 22 (1922) 101.

[2] Wien (WW 211) indicates that he considers this attitude to be the one philosophers naturally take. This also seems to be Moritz Schlick's opinion (*Raum und Zeit in der gegenwärtigen Physik*, 4th ed., Berlin, 1922, p. 92 ff.).

[3] ER 79, 47 (Eng. 412, 385–386). [The English translation, quoting Kant (*Critique of Pure Reason*, B 341, Müller trans.), reads "not a real object."]

[4] Hans Reichenbach, 'La signification philosophique de la théorie de la relativité,' *Revue philosophique* 94 (July–Dec., 1922) 53.

[5] Léon Brunschvicg, *Les étapes de la philosophie mathématique* (Paris, 1912), p. 56.

[6] ER 55, 114 (Eng. 392, 444). [Cassirer quotes Adolf Kneser, *Mathematik und Natur* (Breslau, 1911), p. 13.]

[7] ER 47; Eng. 385–386. [In the last quotation Cassirer is quoting Kant, *Critique of Pure Reason*, B341.]

[8] Spinoza, *Ethics*, Pt. 3, Introduction, and Pt. 4, prop. 4 [*The Chief Works of Benedict de Spinoza*, trans. R. H. M. Elwes (Bohn Library ed.; reprint New York: Dover, 1955), pp. 128, 193].

CHAPTER 19

RELATIVISM AND MECHANISM

161. *Are the Two Theories Opposed?*

Now that we have classified relativism insofar as it can claim to be considered as a philosophical system and have seen it to be a mathematicism, we would at the same time seem to have determined its place as a scientific theory. However, we have yet to discuss its relationship to mechanism, a question that has been much debated since the appearance of Einstein's theory. This discussion seems all the more useful since it has been suggested that there is a complete opposition between the two conceptions of reality we have juxtaposed in the title of this chapter. Such an opposition would corroborate the belief that relativism is entirely different from anything theoretical science had previously imagined, that it perhaps even constitutes a sort of extravagance, a monstrous and no doubt ephemeral excrescence. Our analysis, however, as the reader has seen, tends rather to reintegrate relativism into the framework of scientific thought, to show that relativism conforms to the canons science has followed at all times and in all places, or at least during the periods when it was dominated by the concerns that seem to us to characterize the modern concept of science.

162. *Planck's and Wien's Positions*

Planck, in the work we have so often had occasion to cite, declares that "whoever regards the mechanical conception as a postulate in physical theory cannot be amicably disposed toward the theory of relativity" (PRB 52; Eng. 37); Wien, for his part, believes that this theory makes "the foundations of physics more and more general and indeterminate, which is probably not a disadvantage from the standpoint of a mathematical treatment, but leaves the physicist unsatisfied." In another passage, while admitting that the theory of relativity demonstrates that non-Euclidean geometry lends itself to the representation of natural phenomena, the scientist stresses that "this road leads toward a purely phenomenological conception and can only be followed as long as one is willing to limit oneself to this mathematics" (WW 302, 306); he compares this point of view to that of Mach. We cannot overstate the importance of these opinions; they deserve our closest attention, not only because of the great authority of their authors as scientific theoreticians, but also because, as the reader has seen, both of them strongly favor Einstein's doctrine and are at the same time resolutely hostile to positivistic phenomenalism. Although it is easy to explain the origin of these judgments, we dare say they are largely open to revision.

163. *The Definition of Mechanism*

Needless to say, the way one responds to the question of the relationship between relativity and mechanism depends on how one defines the two terms. There is no doubt about relativity, since it is a recent theory whose author is still alive. Those who aspire to continue Einstein's work proclaim his unique merit (Weyl, STM 2), and if anyone errs in interpreting his work, he is always there to correct the mistakes. But the same is not true, of course, in the case of mechanism, which is a doctrine as old as the world, or at least as old as the attempt to understand the world by scientific reasoning. It is thus only natural that the term sometimes assumes very different meanings as a result of the evolution of the theory throughout the ages, and we shall now try to distinguish between them.

Mechanism in a strict sense can be considered to be the hypothesis that the physical world is composed exclusively of interacting material masses. Consequently mechanism is defined as the explanation of reality in terms of mass and motion. There is yet another distinction to be made, however, concerning the nature of the action one means to allow. For example, one can stipulate that interaction must take place exclusively through contact, by impact, to the exclusion of all action at a distance. The reduction of natural phenomena to such a scheme was certainly the ultimate goal of many modern physicists almost to the end of the last century. It is well-known, however, that they never came close to this ideal. The only really logical attempt to establish an explanatory system with no appeal at all to the concept of force — Hertz's mechanics — remained merely an unfulfilled project, as eminent physicists have noted many times.[1]

164. *Mass and Matter*

On the other hand, one might allow the use of the notion of force, either by positing the existence of simple centers of force like Boscovich, or by a sort of syncretism in which material corpuscles are surrounded by forces somehow attached to them. This is admittedly an explanatory system that at a given epoch (the second half of the last century, to be precise) seemed to the physicists of that time, such as Helmholtz, Lord Kelvin and Cornu, to be very close to being achieved. Nevertheless, one must remember that the concept of mechanism understood in this way is only applicable during this particular period and those immediately preceding it, for explanation by mass and motion is obviously correlative with the definition of the term mass. Now this definition has only gradually emerged in modern physics. Consequently Lucretius, who in so many other respects seems to be the very prototype of a mechanist, is necessarily excluded, for there is no doubt that, although he spoke constantly of impact, most of the time he had only the vaguest idea of the role played by mass in this phenomenon. When he did conceive of its role with some clarity, it was almost certainly in quite a different way from the physicists we have mentioned, since he had no notion whatsoever of the principle of inertia, the foundation of all mechanics. The only way to include Lucretius – and thus reestablish a relationship that is undoubtedly consistent with historical truth and consequently with the true sense of the evolution of science – is by substituting another less precise term for mass, for example by defining mechanism as explanation by *matter* and motion. We shall see in a moment the significance of such a change.

For the time being, let us confine ourself to mass, observing that it is indeed in this sense that the term mechanism is most often used. We have used it this way ourself in the course of this work, in contrasting mechanical explanation with the electrical theory of physical phenomena that was born in the last years of the nineteenth century and reached its height at the beginning of this century.

165. *The Electromagnetic Theory*

It should be noted that mechanics, thus understood, was discredited and one might even say abandoned by physicists in general well before the appearance of the relativistic conceptions. Curiously enough (as we have already discussed, § §41 and 79), the very men who did the most to bring this about were not always clearly aware of the fundamental change taking place. Maxwell, who created the electromagnetic theory of light, is a good example. By reducing all electrical phenomena to his famous equations he laid the foundations for the new hypotheses and at the same time paved the way for the demonstration that these phenomena could not be reduced to classical mechanics. Nevertheless, he used all the resources of his great genius to establish just the opposite:

thet a mechanical explanation was possible in each particular case.[2] Similarly Helmholtz, who first determined the elementary quantity of electricity and was an early supporter of Maxwell's theory of light, retained to the end the hope that electromagnetic phenomena would in the final analysis be recognized as purely mechanical modifications occurring in the ether. Others, however, although they saw the direction in which physics was heading, steadfastly refused to go along. Kirchoff, for example, was fond of stating that the electromagnetic theory could hardly be progress, since it replaced a process that could be represented in a determinate way by something completely indeterminate (because all we know about electricity is that it is propagated in space in the manner of waves, that it is directional and that it produces this or that effect). (Wien, WW 302.)

166. *The Smoothness of the Transition and the Positivistic Explanation*

But, all things considered, this attitude was rather the exception, and in general it is quite remarkable, as Planck aptly points out, "how simply and quietly the transition from the mechanical to the electromagnetic theory was made in physical literature." We might add that this is particularly remarkable, given that it involved (or so it must have seemed) abandoning once and for all the principles constituting the very foundation on which the entire explanatory system of modern physics had been built up to that time. Planck considers that the facility of this change "is a good example of the fact that the kernel of a physical theory is not the observations on which it is built, but the laws to which they give rise" (PRB 137; Eng. 94). But did not this renowned physicist himself teach us, notably in his resounding polemic with Mach, that faith in a reality situated outside the self is, on the contrary, essential and even indispensable to the physicist? "All great physicists have believed in the reality of their representation of the world," he declared, and "contemporary physicists speak the language of realism and not the language of Mach."[3] Did he not also defend – and very effectively – the rights of mechanism as an explanatory theory? And by recognizing the primacy of law over explanation, would he not be accepting the positivistic claims at the very point where they are most unambiguous and, we would say, most divergent from the real direction of scientific thought? Furthermore, in discussing just this aspect of positivism in Chapter 16 (§127), we have shown that both Planck and Wien treated Bohr's theory as a true representation of reality, in spite of its paradoxical nature.

Of course we have also seen (§130) the reasons why the physicist tends to be more or less consciously indulgent toward his chosen theory. But this explanation obviously would not be applicable here, for we are trying to show how the physicist could come to choose the new theory in spite of the fact that it involved a really serious drawback apparently not to be found in the old theory.

167. *The Role of Experiment*

The first reasons to come to mind might be those drawn directly from experiment; certainly experiment has played a quite considerable role in this case, as it has in the history of science as a whole. This role was even more important here because the aprioristic tendency clearly had the opposite effect. Maxwell undoubtedly would have liked, as we have just said, to be able to reduce all of the physics of electricity to a mechanical scheme. Nevertheless, experimental results forced him to formulate his equations as he did, although in his eyes they constituted only a sort of temporary phase and were destined to be explained later in mechanical terms. This was also the attitude of the scientists who followed him; we dare say all of them without exception hoped for such a reduction and for a long time believed it would be possible. They were all surprised, even aggrieved (and this is no exaggeration) at the frightful complications involved in furnishing a mechanical explanation of the most elementary electrical phenomenon, while on the other hand, as Planck points out, "the same phenomena were represented, by means of the Maxwell − Hertz differential equations, very simply and accurately in all details examined up to that time." Thus, in the end, physicists reluctantly arrived at the conclusion that "the assumption of the complete validity of the simple Maxwell − Hertz differential equations excludes any possibility of explaining, on mechanical lines, electro-dynamic phenomena in the ether" (PRB 49, cf. 125; Eng. 35, cf. 85). This conviction was reinforced by the appearance of works directly demonstrating the incompatibility of these two ways of thinking (IR 58; Eng. 64−65). In the meantime the physics of electricity had taken over the immense field of optical phenomena, and experimental results of course played a decisive role in this development as well. It is because he had observed that the velocity of the propagation of light is identical to that of electricity that Maxwell had had the idea of identifying the two fields, and it is the discovery of Hertzian waves, which are electrical in origin but behave like light in all respects, that had won acceptance for the concept. Finally, there came the experiments of J. J. Thomson and Kaufmann suggesting the idea of an identification between mechanical and electrical inertia.

168. *Mechanical and Electrical Inertia*

Still, one is forced to admit that this combination of circumstances is not enough to explain the anomaly pointed out by Planck. Let us examine a little more closely the last phase of the transition we have just summarized, since it is here, as a result of the reduction of the mechanical to the electrical, that mechanism definitively lost all explanatory power. What do the famous experiments of Sir Joseph Thomson and Kaufmann immediately demonstrate? They prove that an increase in the electrical charge of a body can produce phenomena in which the mechanical mass of the body is apparently increased, and that the mass of the electron increases with its velocity, exactly in the theoretically predicted proportion. This leads to the conclusion that the electron possesses only electrical mass and no mass in the ordinary (or mechanical) sense of the term[4]; consequently, a charged body in motion "simulates exactly the familiar mechanical inertia of a lump of ordinary matter," as Sir Oliver Lodge puts it.[5]

But, from the fact that electrical mass or inertia is something analogous to mechanical inertia and can be added to it, it obviously in no way follows that all inertia must be electrical in origin. Nevertheless, this is the conclusion that was drawn from the experiment. How could this have happened? What motivated physicists to make the great logical leap from the affirmation that mechanical mass *could* possibly be considered electrical in origin (which is the most the theory justified) to the affirmation that it *had* to be? The change is still too recent for us to have entirely forgotten how it came about, as sometime happens later when a fundamental concept has finally been definitively established in an area of physics and completely dominates it. Thus we see Berthoud frankly admit, in his excellent exposition of recent theories on the *Structure of Atoms*, that in reality

it is impossible to verify completely the hypothesis that the mass of the electron is entirely due to its electrical charge; this hypothesis was adopted only because it has the great advantage of leading to a very simple notion of matter.[6]

There is certainly much to recommend this opinion, and we shall return later (§203 ff.) to the physicist's tendency to go beyond the limits of the conclusions strictly justified by his experimental observations. The present case, however, would be particularly flagrant, for if we believe Kirchoff, it would end up replacing something completely explained by something just as completely inexplicable.

169. *The Enigma of how Mass Acts*

But this is not at all the case. To understand this, one need only realize what the mechanistic explanation really was, even during the periods when it claimed to be the most rigorous. Mechanism has two elements at its disposal: mass and motion. It is easy to see, however, that they are far from being of equal value as instruments for this kind of explanation, that what could be called the explanatory force of the system resides almost exclusively in the second term. In fact, what we call mass is a coefficient that we presume to be attached to a body and that measures the action the body performs or is capable of performing. Now, in mechanism this action itself remains entirely unexplained. We have stressed this fact many times in our earlier works (cf. IR 61 ff.; Eng. 67 ff.; ES 1:70 ff.) and have also referred to it here (§§41 and 79). It is our ignorance of the nature of this action that has given rise to a number of theories devised to account for it. But none of these hypotheses can stand a logical examination, and so, obviously, mass remains something enigmatic, merely a name. How is it then that it can be used in explanations that seem to satisfy our need for causes? We all know to what extent this was the case. We have already examined this question elsewhere (cf. in particular IR 98 ff.; Eng. 92 ff.), but we shall return to it briefly, expanding on certain points.

170. *The Familiar*

The most widely held opinion makes use of the notion of the *familiar*. It was clearly formulated by John Stuart Mill,[7] and Mach has often expounded it quite forcefully. Now, the very fact of the change with which we are dealing here clearly contradicts this point of view. If the purpose of explanation were to reduce less familiar phenomena to more familiar ones, no one would have been able to take a concept such as the electromagnetic theory of matter seriously, since it reduces one form of action – the only one we constantly use – to another form that makes use of a force beyond the reach of our immediate perception.

171. *Touch and Action at a Distance*

It must be recognized, however, that at the beginning, when the rudiments of our notion of corporeity were being formed, the fact that this notion seems to be furnished ready-made, so to speak, by the immediate sensation of touch must have played a considerable role. Nevertheless, one should take note of several essential points.

First, let us note that it seems altogether useless — although not uncommon — to call upon the sense of touch to explain why we find explanations in terms of action by contact preferable to those explanations based on action at a distance. Indeed, all one need bear in mind is that, if we are to understand anything about an action occurring in time and space, we must consider time and space to be continua and the action in them to take place continuously. It would seem inconceivable to us for one phenomenon to result from another if they were separated by any measurable lapse of time without anything at all happening during that time; every phenomenon must be the consequence of what immediately precedes it in time. Likewise, we cannot understand how an action can take place without anything happening in the intervening space, whereas it seems natural if the bodies are touching. This is really only a negative condition, however, or rather, to use terms familiar to the mathematician, a necessary but not a sufficient condition, for it does not show us why action by contact is apt not only to seem less inexplicable than action at a distance, but even to seem genuinely understandable in itself. In order to see this we must submit the concept of touch itself to closer analysis.

172. Touch-As-Sensation and Touch-As-External-Phenomenon

One immediately recognizes that this word, whose meaning seems so clear at first glance, actually unites two different notions, namely, touch-as-sensation and touch-as-external-phenomenon. This union and the resultant ambiguity are clearly what is behind the claims concerning the *primacy of touch*, according to which this sense is considered particularly important for indicating materiality. When Lucretius proclaims that "no thing can touch or be touched except body,"[8] it must be assumed that he supposes that the *thing* can touch; consequently his statement cannot refer to touch-as-sensation without changing the meaning of the word 'touch'. Furthermore, this statement is preceded by a series of verses where the poet-philosopher gives a magnificent description of the destructive effects of a tempest, concluding: "Winds have invisible bodies since in their actions and behavior they are found to rival great rivers, whose bodies are plain to see." What he is in fact doing is demonstrating the materiality of air, and for him the criterion of materiality is clearly the ability to produce an effect. Let us also recall that at the beginning of this work (§4) we recognized with Bergson the exclusively qualitative nature of the 'immediate data' of our consciousness. It is certain that the data furnished by the sense of touch are no exception in this respect and thus can contain nothing of the quantitative spatiality that appears to us to be inextricably bound up with the concept of body as it exists in common sense perception.

But although touch-as-sensation has no advantage over the impressions of the other senses, it is certainly the sense that gives us the illusion of comprehensibility, or in any case is an indispensable component of this illusion. However, in order to understand this better, we must appeal to yet another distinction.

173. *The Feeling of Volition*

The ability to produce an effect of which we just spoke is something we feel in ourselves; we sense in an immediate way that our muscles obey our will and that we can act on material bodies by moving our arms and legs. Nevertheless, these sensations are not at all like the others, precisely because of the intervention of our will. They do not belong (in the words of Schopenhauer, who is not above criticism on this point) uniquely to the world of representation but also to the world of the will. But we must not lose sight of the fact that Schopenhauer was certainly wrong not to have laid more stress on the distinction between the two worlds. This is a profound distinction, for while I am able to conceive of the world as *my* representation in the sense that it has its roots only in my sensation, it is absolutely impossible for me to conceive of the world as *my* will. Maine de Biran saw this quite clearly when he asserted that it is by the will and by that which opposes it (which can be presumed to be analogous to it) that we arrive at the conception of an external reality.[9] Without going that far, we can grant that the concept of an external world, whatever its origin, is confirmed in this way, and this helps us better understand the important role played by touch-as-sensation in our concept of materiality.

174. *The Enigma of the Act of Throwing and its Explanation*

It is certain, on the other hand, that, at least insofar as common sense is concerned — that is, before our reasoning is tainted by philosophical doubt — the effect we produce by continuous contact seems entirely comprehensible. It could be claimed here again that this is only because the action is so familiar. But it is here that a precise example enables us to understand why the concept of familiarity is insufficient. Indeed, since the origins of humanity one of our most familiar acts has certainly been the *act of throwing*: even apes know how to throw. As we saw in §38, far from being taken for granted, the behavior of the projectile, the fact that it continues to move after the hand has ceased to touch it, seemed as late as Galileo to be so enigmatic that theories as extravagant as Aristotle's were called into play to explain it. Moreover, there could be no question here of a conviction based on external experience, because external experience shows us, on the contrary, that motion produced by the impact of inanimate bodies is just as persistent after contact ceases as the motion produced by throwing. Of course one could call upon a mysterious tendency toward unity, claiming that we feel the need to reduce all action to a single type; as a result, however familiar throwing may be to us, motion by permanent contact is even more so, and this is why we attempted to reduce the first to the second. Who cannot see how contrived such a theory is, given the intense feeling of mystery Aristotle (and with him all of science) seems to experience in the face of this phenomenon? Therefore, one is more or less forced to conclude that we are dealing here with the simple result of a transposition from the subjective to the objective. If Aristotle judges the action caused by contact to be natural and immediately comprehensible as long as the contact lasts, it is because he is thinking of the action performed by the subject and the accompanying tactile sensations; when they cease, the action ceases to be understood and should therefore stop. Aristotle's theory, then, is actually an attempt to reconcile with the experience of motion a conception of body that is inherently familiar to us because it corresponds to a whole set of considerations in which touch-as-sensation and its confusion with touch-as-external-phenomenon play a considerable role, since the fact (or the illusion, if one prefers) of comprehensibility does not hinge on a unique and privileged sensation, but results from a combination of external sensations with other internal sensations depending on the exercising of the will.

But mechanics is not able to retain this point of view; on the contrary, it had to resign itself (after a resistance that was, therefore, all things considered, understandable) to replacing 'environmental reaction' by *vis impressa* and finally by the concept of inertia. As a result, since the body moves without remaining in contact with anything at all, the ability to produce an external effect, such as we see it explained by Lucretius, becomes dominant. As for the continuation of motion, we henceforth understand it as the conservation of velocity, which ends up changing the way we understand the relationship between space and the material bodies it contains (as we saw in §37).

175. *Mass as Substance*

Physics immediately goes further. Indeed, as soon as we consider that velocity must be conserved in the body and consequently have become accustomed to treating velocity as a substance, does this not suggest that velocity, or at least something we can consider *equivalent* to it (this is the first thing to which we resort whenever strict identity escapes us) will be conserved in passing from one body to another? And this is why we see the rudiments of the concept of the conservation of energy — namely, the Cartesian principle of the conservation of motion — appear simultaneously with the principle of inertia and finally, toward the middle of the last century, assume the definitive form under which it has been universally accepted. The ability to move other bodies by impact is measured by a coefficient called mass, which henceforth becomes the sole characteristic of a body.

176. *The Link with Touch-As-Sensation*

Nevertheless, despite the precision of these notions on which mechanism in the strict sense (discussed in §163) is founded, the fact remains that the way things are set in motion is by the *impact* of bodies. This fact alone is sufficient to remind us constantly of the older concept of the material and thereby to reestablish the link with touch-as-sensation. This happens all the more easily because the older concept is that of common sense, and as such is therefore never completely eliminated from our thoughts. On the contrary, we use it continually (as we do everything that is part of this ontology) in countless circumstances of our daily life. " 'Tis certain," says Hume,

that almost all mankind, and even philosophers themselves, for the greatest part of their lives, take their perceptions to be their only objects, and suppose, that the very being, which is intimately present to the mind, is the real body or material existence. 'Tis also certain, that this very perception or object is suppos'd to have a continu'd uninterrupted being, and neither to be annihilated by our absence, nor to be brought into existence by our presence.[10]

For this reason, Bergson is correct in observing that atoms said to be deprived of physical qualities are in reality defined "only in relation to an eventual vision and an eventual contact."[11] And even the ether — no matter how remote it seems from our sensation and no matter what pains physicists take to distinguish it from matter by attributing to it extravagant (and in the end contradictory) properties that no real matter could possess — retains, by the sole fact that it is supposed to exercise mechanical action (given that it was invented for the sake of this action), something of the material and thereby a connection, however distant, with the sense of touch.

177. *It is Broken in Electrical Theory*

This link is finally broken in electrical theory. Of course, like all scientific theories, this one begins with certain facts of sensation and tends to come back to them to some extent, since it must in the final analysis explain these very facts of sensation. But in this case the data, which are those of the phenomena of the electrical field, are so far from tactile sensations that when we think about these data we evoke neither sensations of touch nor those associated with the exercise of our will. Thus the mechanical phenomenon, in spite of the mystery involved in action by impact discussed in §41, gives us a certain illusion of comprehensibility, an illusion so strong that not only have physicists until fairly recently affirmed the complete rationality of this phenomenon (as we have indicated), but philosophers such as Avenarius and Hermann Cohen were also able to dispute the validity of Hume's nevertheless quite decisive argument.[12] The electrical phenomenon, on the contrary, clearly appears to us as something unknown or even unknowable. This is the real source of Kirchoff's resistance to the electrical theory of light.

178. *Electrical Theory and Anthropomorphism*

But we have seen that this resistance was isolated, and most physicists passed without the slightest qualm from Fresnel's theory to that of Maxwell and Hertz, then went on to accept the theory that matter itself is only an electrical phenomenon. Although this change may seem profound and revolutionary, it was in fact carried out in accordance with the general scheme that governs the progression of science. Indeed, by showing how the concept of material body, first formed in strict dependence on our sensations of touch and volition, gradually frees itself from these ties, all we have done is illustrate, by use of a particularly precise and important example, one aspect of Max Planck's apt observation that science, as it progresses, moves further and further from "anthropomorphism" (§22).

179. *Explanation by Substance*

Throughout the evolution from common sense to the electrical theory of matter, does there remain no link between the new concept and the old ones?

On the contrary, it is obvious that one extremely significant fact does remain: in each of the phases we have just discussed, the explanation hinges on the persistence of something, of a substance, through space and time. This eternal framework of any system of explanation was suggested by Aristotle when he observed that the atomism of Leucippus and Democritus had developed directly from the Eleatics, retaining their immutable being, merely multiplying it in order to be able to 'save' diversity in time and space by use of the concept of motion (cf. IR 101 ff.; Eng. 94 ff.).

180. *Mechanism as Explanation by Motion*

Consequently, if one wishes to reduce the definition of mechanism to its barest essentials, one must disregard the notion of mass and limit oneself to stipulating that explanation will be in terms of motion, it being understood that that which moves and which can only be an indeterminate principle of action will have to remain identical with itself; it will be a *substance*, to use a metaphysical term. The substitution of the less precise term matter for the term mass (mentioned above, §164) is clearly a step in this direction.

Furthermore, it is not impossible to find similar pronouncements by physicists. Significantly enough, Planck himself sometimes defined mechanism in a way allowing of such an interpretation. In fact, at the very beginning of the lecture from which we cited the statement on the conflict between mechanism and relativity, we read that mechanism is the conception that all physical phenomena can "be completely reduced to movements of invariable and similar particles or elements of mass." A few lines later he adds that according to this conception, "the problem for theoretical physics is simply to interpret all phenomena in Nature in terms of motion" (PRB 40; Eng. 28). By understanding mechanism in the sense in which it is used in this last sentence, we shall have the great advantage of not separating what forms a clear unity from the historical point of view, for the close connection between Democritus and Descartes, just as between Boyle and Boscovitch, will be immediately apparent. We shall also understand that the electromagnetic explanation, in spite of the considerable differences that seem to separate it from the mechanism of mass, nevertheless belongs to the same family. And this is certainly the principal reason it was possible to make an all but imperceptible transition from one theory to the other.

181. *How Relativism Resembles Mechanism*

If we now turn again to the relativistic conception, we shall see that it stands in essentially the same relation to mechanism as does electrical theory. To be sure, in its final phase relativism makes matter disappear; but it must be remembered that, understood as a substratum producing a purely mechanical effect, matter had already disappeared earlier. And there does still remain in the new theory something, a *substance*, that is immutable and creates phenomena by its motion — because matter is made of energy, which is supposed to be conserved.

What moves in this way is actually even harder to describe in ordinary language than was the element in electrical theory, but this is clearly only a matter of degree, for the latter was already entirely foreign to our sensation. It is also true that the immutability of the element that moves in relativity theory does not extend to its coefficient of mechanical action, that is, to its mass, because mass, on the contrary, is supposed to vary with velocity. But here again, as we have seen, this is something relativism inherited from electrical theory, which also assumes, independently of any consideration having to do with relativity, that the mass of an atom must be a bit less than the sum of the masses of the positive and negative electrons of which it is composed because of the proximity of the charged particles to one another.[13]

Finally, if we consider things from the most abstract viewpoint, which is that of spatial explanation in its broadest sense, it is certain that Einstein's theory and those developing out of it do not depart from the guidelines followed by scientific explanation up to that time. For mechanism does not limit itself to accounting for diversity in time; it also intends to explain simultaneous diversity. This intention is what is behind the concept of the unity of matter in particular — mechanism's 'secret postulate' — as well as the countless attempts to connect matter to the ether, that is, to physical space. These attempts to define matter in terms of the ether, as ring vortices, etc., characterized precisely that period of the nineteenth century we mentioned above, when mechanical theory was at its height. And relativism, by absorbing all of physical reality into space, is only completing the work undertaken by its predecessor.

Thus, the more closely we look, the more the alleged opposition between relativism and mechanism tends to diminish and almost disappear.

182. *Why they were Thought to be Opposed*

But how can we explain that men as eminent as those we have cited, men not only fully competent to speak in the name of the emerging science but also deeply imbued with its philosophical principles, could mistakenly believe that such a conflict exists and thereby attribute to relativism a phenomenalistic attitude it clearly does not have? In our opinion, the explanation depends entirely on the opposition pointed out in Chapter 17 (§145) between the vigorously realistic thought of the physicist and the ultimate goal of his science, which is necessarily idealistic since it aims at an explanation of the whole, a deduction of the whole from the content of reason. This opposition is inherent in all science, of course, but it is in some sense hidden, and the physicist is therefore able to ignore it. Relativism, on the other hand, because it is a highly synthetic conception, that is, a genuine system of universal deduction of reality, exhibits this contradiction much more openly and explicitly. Therefore, the physicist's mind is more likely to be disturbed by it and to take refuge in the assumption that the doctrine is fundamentally different in this respect from doctrines science had previously elaborated. From this point of view, then, the state of mind manifested by Planck and Wien is closely related to that of Weyl and Eddington when they assume that behind the reality of relativism there is another reality even more real. But although their assertions are perfectly understandable from the psychological point of view, we find it difficult to justify them as following from the nature of the theory itself.

183. *Acosmism and the Positivistic Illusion*

This is also the element foreign to the pure and simple doctrine of Auguste Comte to which we alluded in Chapter 5 (§52) in connection with statements by Vouillemin and Moch. Indeed, the example of Mach shows us how much affinity there is between absolute phenomenalism and positivism. The fact that acosmism, the disappearance of reality – which is necessarily the position at which all physics ultimately arrives – becomes so pronounced in the relativistic conception might lead to the assumption that here one can dispense with all reasoning about reality. Moreover, we have already seen that Planck, whose position usually diverges widely from Mach's nevertheless tends to approach Mach's position to some extent in his views on relativism. But though more refined than the positivistic position, Planck's argument, as we have demonstrated, is basically no more legitimate. In relativism, as in all physics, acosmism is found only at the extreme limit, while the entire body of the doctrine is resolutely realistic and consequently entirely analogous to the explanatory theories that preceded it, and to mechanism in particular.

NOTES

[1] Cf., for example, Max Abraham, 'Die Neue Mechanik,' *Scientia* 15 (1914) 8.

[2] Henri Poincaré, *Electricité et Optique* (Paris, 1901), p. 3 ff.

[3] Max Planck, *Die Einheit des physikalischen Weltbildes* (Leipzig, 1909), pp. 36, 37. We shall have occasion to return to this polemic later (§ 271).

[4] WW 11 ff. J. J. Thomson's experiments date from 1887 and Kaufmann's from 1891.

[5] Sir Oliver Lodge, 'On Electrons,' *Journal of the Institution of Electrical Engineers* 32 (1902–1903) 50 [Meyerson quotes *Sur les électrons*, trans. Nugues et Péridier (Paris, 1906), p. 14].

[6] A. Berthoud, *La constitution des atomes* (Paris, 1922), p.15. A very clear account of how matter is somehow reabsorbed into the electromagnetic ether can be found in Louis Rougier (*La matérialisation de l'énergie*, Paris, 1919, p. 62 ff.).

[7] John Stuart Mill, *A System of Logic* (London, 1884), p. 369.

[8] Lucretius, *De Natura Rerum*, Bk. 1, Vol. 305 [*On the Nature of the Universe*, trans. Ronald Latham (Baltimore: Penguin Books, 1951), p. 36].

[9] Maine de Biran, *Science et psychologie*, ed. Alexis Bertrand (Paris, 1887), p. 11.

[10] David Hume, *A Treatise of Human Nature*, ed. L. A. Selby-Bigge (Oxford: Clarendon Press, 1888), Bk. 1, Pt. 4, §2, pp. 206–207.

[11] Henri Bergson, *Matière et mémoire* (Paris, 1903), p. 22 [*Matter and Memory*, trans. Nancy Margaret Paul and W. Scott Palmer (London: George Allen & Unwin, 1911), p. 26].

[12] Cf. Harald Höffding, *Der Totalitätsbegriff* (Leipzig, 1917), p. 78.

[13] Cf. Sir Ernest Rutherford, 'The Electrical Structure of Matter,' *The Times*, 13 Sept. 1923, p. 16, col. 2 [Rutherford has just explained that "we know that the hydrogen atom is the lightest of all atoms, and is presumably the simplest in structure, and that the charged hydrogen atom, which we shall see is to be regarded as the hydrogen nucleus, carries a unit positive charge. It is thus natural to suppose that the hydrogen nucleus is the atom of positive electricity, or positive electron, analogous to the negative electron, but differing from it in mass"].

CHAPTER 20

RATIONAL EXPLANATION
AND THE PROGRESS OF MATHEMATICS

184. *The Transformation of the Notion of Space*

In order to make reality rational, the relativistic conception – like all modern theoretical physics – makes use of mathematical deduction. Now, mathematics does not stand still. It progresses, sometimes with astonishing rapidity – as during the relatively short period of time that saw the discoveries of Descartes, Fermat, Newton, and Leibniz – sometimes more slowly, but it goes without saying that it is constantly progressing and changing. It is therefore legitimate to ask what influence this progress has on the problem of explanation, which is the fundamental problem of all physical science.

Relativism, by the very fact that it is an extremely advanced form of the mathematization of reality, seems to offer a good standpoint from which to study this important question.

From what we have seen concerning the role of the spatial in Chapter 3, it can be concluded that the principal repercussion of the progress of mathematics in the field of scientific explanation was necessarily a transformation of our notion of space. Therefore, it is by taking a closer look at the forms of spatial explanation that we shall be most likely to understand the evolution of this influence.

We called attention in §29 to the fact that among the alternative modes of spatial explanation we had enumerated in an earlier work, only explanation by motion is peculiar to modern physics. Then, examining the origin of the principle of inertia more closely, we recognized that, insofar as it is an explanatory theory, it rests on a profound transformation of the notion of space (§37).

Was this transformation due to some new development in mathematics? We think not.

185. *Mathematics and the Origin of the Principle of Inertia*

Brunschvicg has said that with the modern concept of motion (as seen in Descartes) "there appears a form of intelligence that replaces a different form of intelligence to which it bears no relationship" (EH 185). Without venturing quite that far, we would be inclined to agree with this opinion, provided that it is understood that the change did not result from progress achieved in mathematics.

To see that this is true, one need only examine the way in which the principle is presented, by Galileo as well as by Descartes and Newton. As we know, Galileo does not clearly spell out the principle of inertia; he leaves it implicit. But where he uses it, there is not the slightest reference to any sort of mathematical deduction. Now this is surely not because he in any way failed to appreciate the general influence of mathematics in physics; on the contrary, he explicitly affirmed that it was the predominant element in our knowledge of reality. "I truly think," he writes toward the end of his life, more or less summarizing the principles that had guided him in his research, "that the book of philosophy is the book of nature, which is perpetually open before our eyes but, because it is written in letters different from those of our alphabet, cannot be read by everyone. The letters in this book are triangles, squares, circles, spheres, cones, pyramids and other mathematical figures quite appropriate for this kind of reading."[1] Given that the ideas of the great Florentine have this *panmathematical* foundation, it is all the more significant that he made no use of mathematical deduction in connection with the concept of inertia, despite the fact it was so new and so difficult to establish. And it is equally significant that Descartes, whose great mathematical genius is well-known and who, by creating analytic geometry, had just transformed the whole shape of this science, brought nothing mathematical into his exposition of the principle: in both *Le Monde* (cited in §32) and the *Principles* his deduction is purely philosophical. The same is true for Newton, who nowhere connects the concept of inertia with his great discovery of the infinitesimal calculus, nor with any mathematical deductions at all. D'Alembert did present a quasi-geometric demonstration, but for him the mathematical apparatus is at bottom only a sort of pretence and super-fluity (cf. IR 126 ff.; Eng. 121 ff.). Moreover, it is not clear that d'Alembert's demonstration had any influence whatsoever on the course of science.

But although the progress of mathematics was not the determining factor in the development of the modern concept of motion out of the principle of inertia, it would be a mistake to claim that it had no influence on this concept.

The best way to see this is to go back to the paradoxes of Zeno of Elea. What these admirable arguments make especially clear is how difficult it is for our mind to grasp the continuous. There is no doubt, as Bergson showed in many passages which have justifiably become classics, that it is the discrete, the discontinuous, that immediately seems to conform to our reason, to be rational; anything not discontinuous we try to force into conformity. Nevertheless, it is easy to see that the Eleatic *reductiones ad absurdum* appear much less disturbing to us today than they did to the ancients. Thus, even while occupying a determinate place, the *Arrow* still seems to us to have retained a *velocity*, and therefore it does not seem obvious that it must remain where it is. Where do we get this way of seeing things, obviously unknown to the ancients? There is no doubt that it came from the principle of inertia, which has taught us that velocity is something that is conserved. But it was reinforced by the fact that we customarily posit a velocity at each particular point of any motion whatsoever, that is, we somehow consider motion continuous and discontinuous at one and the same time. In other words, we understand motion as continuous by means of deductions that seem at first to decompose it into discontinuous fragments and then to melt them together in order to reestablish continuity. What we have just characterized in such a circuitous way is, needless to say, the principles of infinitesimal calculus, and this calculus, by the mental habits it has created, has therefore certainly contributed to the establishment of the modern concept of motion. But it did not engender it. On the contrary, as is easy to see from even a cursory study of the writings of not only Newton and Leibniz but their predecessors as well, it is ideas about motion — understood in the Cartesian sense, of course — that assisted in the birth of the calculus.[2] This was a case of action and reaction: the new method, once created, confirmed the concepts that had contributed to its creation.

186. *Reason, the Continuous and the Discontinuous*

Here, however, we must digress to meet an objection that could not fail to have occurred to the reader. Throughout this work we have assumed that reason seeks to understand physical discontinuity by reducing it to spatial continuity. But now we are maintaining that in mathematics, on the contrary, it is the discontinuous, the discrete, that seems to be rational, and that it is by means of the discrete that reason seeks to grasp the geometric continuum. Is there a contradiction between these two ways of understanding? Of course there is, but it is the fundamental contradiction inherent in reason itself. Reason, as Plato said, cannot act, cannot reason, unless it starts from something sensible. But the sensible — which in physics is matter as known by common sense and in mathematics is space as furnished by our immediate intuition — precisely because it is sensible then necessarily appears to reason as foreign to its own nature. Reason has only one way of explaining what does not originate within itself, which is to reduce it to nothingness. And this is why it understands the physical by absorbing it into undifferentiated space, and understands the spatial, the geometrical, by reconstituting it out of points deprived of all spatial dimension. Although these two manifestations appear to conflict, they actually arise from a single fundamental tendency. Having dealt with this, let us return to our subject.

187. *Mathematics and the Origin of Relativism*

In the development we have attempted to retrace, it is particularly interesting to note that the only way infinitesimal calculus had the effect we have described was by influencing our ideas concerning motion in space, which is to say, our idea of the nature of space itself. But in order to avoid any misunderstanding, let us be more precise: any method of calculation can always be used for the simple description of phenomena, that is, in our terminology, for a science concerned only with laws. Our restriction refers only to causal or theoretical science, that is, science that attempts to explain phenomena.

In Chapter 6 (§64) we observed how much the Einsteinian explanation of gravitation resembles the way Galileo and Descartes explained the motion of throwing. However, there comes a point at which the analogy between what occurred in the seventeenth century and what we are witnessing today breaks down. For unlike what happened then, this time the mathematics preceded the physics. Not only is the concept of hyperspace in fact much older than Einstein's theories – so much so that, as Einstein himself remarked, all the work of mathematical development the new theory needed was in a sense ready and waiting when the theory emerged[3] – but prior to the appearance of relativity theory there had already been many attempts to use the fourth dimension for physical explanations.[4]

188. *The Role of Analysis*

Let us also note that what we said above about the influence of the progress of mathematical knowledge in general – that is, even that unrelated to geometry and space – is verified here as well. Indeed this influence is obvious in the present case. One can, of course, arrive at a conception of a non-Euclidean geometry by strictly geometrical means: this is how Lobachevsky (who is, as we know, the founder of this whole current of modern thought) developed such a concept, simply by considering the postulate of parallels.

But it would seem impossible to deny that, following this geometrical development, purely analytic considerations played a leading part. To express this idea in its most elementary form, it seems obvious that the very idea of a fourth dimension is suggested to us by a comparison between geometry, where the number of powers is limited to three, and algebra, where the number of powers is unlimited. And similarly, all the subsequent work that made this concept familiar to mathematicians and, as we have just seen, established the context for the new theories, was in the field of analysis. Finally, we know how much relativity owes to Minkowski's formula, where time is represented by an imaginary quantity, a notion that is certainly purely analytic.

Thus, Brunschvicg quite appropriately stresses the fact that "in order to appreciate fully the service rendered to science by the instrument of mathematics," it "is necessary above all not to limit abstract analysis to the arithmetic of Pythagoras, as Kant did implicitly or Neo-Criticism explicitly" (EH 591–592). Indeed, the significant thing about this evolution is that it is due to the *progress* of mathematics. But it is necessary to add what is for us an essential point: in the final analysis we are dealing with a geometry, and it is because the Einsteinian physicist has modified his notion of space that he *understands* gravitation. In short, from the standpoint of making reality rational, everything derived strictly from analysis has borne fruit only because it has been applied to the notion of space and change in space.

NOTES

[1] Galileo, 'Letter to Licati' (1641), *OEuvres*, ed. Alberti (Florence, 1842), 7:355.

[2] Wien is only reminding us of universally accepted facts when he observes that "not only have differential and integral calculus essentially grown out of problems of physics, but in addition, differential equations, Fourier series and potential theory are, in the final analysis, physical in origin" (WW 18).

[3] Cf. ES 2:378. Weyl (STM 77 ff.) provides an excellent historical account of non-Euclidean geometrical theories, pointing out that as far back as antiquity Proclus already had doubts about the theorem [sic] of parallels. Eddington, in his turn, stressed the importance of the mathematical theory of tensors for the emergence of the concept of relativity, although Riemann, Christoffel, Ricci and Levi-Civita had never thought of applying their calculus to gravitation at the time when they developed it (STG 2–3; cf. p. 212 and Eddington, *Espace, temps et gravitation*, trans. J. Rossignol, Paris, 1921, Pt. 2, pp. 40, 53). [Eddington provided a second, mathematical part for the French edition, not included in the English edition. For very similar material, see Eddington, *The Mathematical Theory of Relativity* (Cambridge, 1923), pp. 58–59 (§27) and pp. 69–70 (of §33).]

[4] ES 1:160, 2:444. Weyl recalls that Riemann had rightly anticipated that theories of

hyperspace could someday be used to account for facts the Newtonian concept was unable to explain, stressing the importance of "preventing the work from being hampered by too narrow views" and "keeping progress in the knowledge of the inter-connections of things from being checked by traditional prejudices." Weyl adds that Riemann's prediction was realized seventy years later by Einstein (STM 97, 102).

CHAPTER 21

PROGRESS IN MAKING THINGS RATIONAL

189. *Transitive Action*

If we now survey physical reality as a whole in an attempt to ascertain what aspect of it is made rational by the use of the geometry of hyperspace, it would seem readily apparent that we are entering an area that was justifiably attributed to the irrational. Indeed, the enigma of gravitation was obviously only a particular form of the more general mystery of transitive action, the action one body is capable of exercising on another. Since, on the other hand, everything other than gravity is supposed to be capable of being reduced to electrical phenomena, we consequently arrive — granting, of course, that Einstein's continuators have succeeded or will succeed in resolving this problem — at the possibility that, contrary to what Hume's demonstration might seem to establish, the whole class of transitive phenomena will be entirely explained, brought into conformity with the canons of reason.

Clearly this is an extraordinarily remarkable result. But perhaps we shall better understand its importance if we consider it from a somewhat more general point of view.

190. *Becoming*

Transitive action is only one aspect of *becoming*, the continual change that seems to be the true essence of the real. It is because there is change and we feel the need to explain it that we arrive at the notion of cause, and the entire business of our thought — of which science is only one particular area — revolves around this problem. Reason attacks the problem from different sides, its preferred method being to seek to demonstrate that change is only apparent, that it hides a deeper identity — these are the notions of conservation — and then, wherever this notion of identity is found to be in too flagrant contradiction with the facts, to try to bend the rhythm of thought to the flux of phenomena. As we have seen, one can distinguish two principal ways of proceeding: mathematical deduction and what we have termed (for lack of a better word that is as comprehensive) extramathematical deduction. The former method is the one followed by modern science, and it is thus by

176

means of mathematical deduction that one aspect of the problem of becoming has been made rational. But this advance is not the only one of its kind that modern physics has to its credit. For, prior to relativism, kinetic theory, using Carnot's principle, attacked a different side of the problem with at least as much success. In fact, with this concept, through the genius of Gibbs, Maxwell and Boltzmann, the continuous and by nature irreversible change presented to us by the whole of reality — a change Carnot's principle captures in a precise formula — is explained by reducing it to a change in the position of elements assumed to be inalterable, a change that can be established as logically necessary by using the notion of probability combined with that of the conservation of motion. And it is in an analogous manner that Einsteinian relativity attempts to make gravitational motion comprehensible, that is, rationally necessary.

191. *Things Cannot be Made Completely Rational*

Obviously things can never really be made completely rational. To explain irreversibility, an improbable state at the beginning of time must be accepted as given. To explain transitive action there are several givens of the same sort that relativity theory cannot claim to have made rational: on the one hand, there is that of spatial discontinuity (Chapter 13), and on the other, that of interpretation (Chapter 14). But perhaps what most resembles the given of the 'improbable state' is the notion of existence, as distinct from essence, which we discussed in Chapter 12. For it seems quite clear that we must abandon the possibility of ever deducing this existence from anything at all.

192. *Gravitation and Action by Contact*

Furthermore, let us call attention to a surprising thing about gravitation: although at first glance is certainly seems to be due to action at a distance, it can be made rational more readily than can action by contact. This is the direct result of the fact that, by virtue of the prevailing hypotheses, action by contact is dependent on electrical forces. Even the staunchest supporters of Weyl's and Eddington's theories (including the authors of the theories themselves) will certainly not go so far as to claim that these theories are worked out as fully as Einstein's general theory of relativity. Clearly there is something here that is contrary to our most unquestioned immediate intuition, for which action by contact is perfectly familiar while action at a distance seems to be a genuine enigma (§60).

193. *Gravitation is Transmitted Without any Intermediary*

It should also be noted that, contrary to what certain writings would seem
to imply, the property of action at a distance does not totally disappear
from gravitation in relativity theory. Gravitation is not transmitted instantaneously, but with a known and limited velocity, the velocity of light. It
is an action quite similar to the one Faraday supposed to be the determining
cause of electrical phenomena. In relativity the gravitational motion of a
body is seen as depending on the local state of the spatio-temporal universe
in the immediate neighborhood of the body submitted to the action, just
as for Faraday the motion of the electrified body depends on the state of
the neighboring dielectric. There is a difference, however: the relativist
involves space itself in the process, or, if one prefers, the particular structure
of this space, since the action is transmitted without the intervention of
any material intermediary, such as a gravitational ether. If, contrary to
all possibility, a mass could suddenly be created in a given part of space,
the other masses, as soon as they were affected by its action, would obey
it — without there being any radiation of *force* between them, as the Newtonian theory more or less explicitly posited — by the simple fact of a modification of space, a *curvature* equivalent to what appears to us as mass.[1]
It is really quite extraordinary that we should come to understand in this
way something that originally seemed so paradoxical. But we can recall
here, as we showed in §64, that as a matter of fact, for our immediate
intuition, inertial motion is also a phenomenon of change — a passing shot
does indeed seem to change its position continuously — and that, by virtue
of the Cartesian principle, we nevertheless have no difficulty understanding
that this change is necessitated by the very nature of space.

At any rate, here once again, in the case of the relationship between gravitation and action at a distance, we find a refutation of the old apothegm, too
often cited as a guiding principle for all scientific research, that the more
familiar phenomenon is the more comprehensible.

NOTE

[1] "From the physical point of view, Mach's theory of relativity, as developed by Einstein, can be interpreted as a complete return to the doctrine of action at a distance
if one defines action at a distance as an action having no intermediary other than space
....According to relativity theory, all forces acting across empty space can only be
considered forces acting at a distance. Should it be necessary for these forces to be
propagated with a finite velocity, specifically the velocity of light, this fact constitutes
a new difficulty for the theory, given that one cannot conceive how it is possible to
attribute to empty space a physical property such as the presence of a velocity, namely,
the velocity of light" (WW 306, 307).

CHAPTER 22

THE APRIORISTIC TENDENCY AND EXPERIENCE

194. *Was Descartes the Victim of an Illusion?*

In spite of everything, it is certain that many scientists (as we were able to verify personally during an interview with one of the most distinguished mathematicians of our time) are likely to be shocked when someone tries to demonstrate the strict analogy between the modern system and that of Descartes, an analogy we developed in Chapters 10 and 11. Is this reaction entirely unjustified? Not at all. It stems from the fact that the relativist is aware of the formidable obstacles the theory had to overcome, of the really extraordinary audacity Einstein needed to conceive the paradoxical idea of assimilating inertia and gravitation, of the wealth of ingenuity his mathematical genius had to display in order to achieve this assimilation, starting from the accumulated results of some of the most illustrious mathematicians who had preceded him and then employing dazzling feats of analysis. He knows too what a prodigious number of efforts nineteenth century physicists had to bring into play to unify what used to be called the diverse forces of nature, thus permitting Weyl and Eddington to attempt a geometrical deduction of the whole of physical reality by the reduction of electrical phenomena to space. How could Descartes, no matter how great his mathematical genius, pretend to arrive at an analogous result with only the resources of quite elementary analysis at his disposal?[1] Is it not more likely that he gloried in an altogether chimerical accomplishment, that he was, in short, the victim of a pure and simple illusion?

In reply, we must first point out that if there was an illusion, it was singularly strong and durable. Not only was the powerful mind of Descartes himself completely taken in by it, but for more than a generation, until the great works of Newton, European scientific opinion embraced it without reservation.[2] And in spite of the brilliant success of the Newtonian concepts in the area of physical and especially astronomical explanation, great effort and many long years were required before they prevailed over the Cartesian system. This external evidence alone would make us suspect that something more profound and essential than an illusion was involved. It seems to us that the principles that have guided us in our analyses of the phenomena of scientific evolution enable us to see this with some clarity.

179

195. *The Distinction between Science and Philosophy*

In Chapter 18 of our earlier work (ES 2:354 ff.) we explained how science, though seeking the same goal as philosophy in the sense that it too attempts to make the whole of reality appear to depend upon thought itself — that is, to be necessary — differs from philosophy in that the reasoning that creates it and is manifest in it — scientific reasoning — accepts, at least provisionally, the existence of the irrational coming from outside, from perception. For philosophy, on the other hand, such an attitude would constitute a sort of suicide, to use Burnet's apt expression. It is this characteristic resignation or acquiescence that enables science to stay constantly close to reality in its reasoning and consequently allows experiment to play such an important role. But this resignation is only temporary, for reason never waives its rights. After apparently accepting the perceptual data that constitute its point of departure, science then turns against these data, attempting to explain them by means of its reasoning and its theories, and does not stop until they have been absorbed into space. Thus the real distinction between the two is that science seeks to attain gradually and indirectly what philosophy believes it can achieve all at once. But the difference also lies even more markedly in the fact that science, unlike philosophy, proceeds unconsciously. The scientist does not know the true goal of his deductions; on the contrary, he remains firmly convinced that, far from aiming at the destruction of reality, he is constantly confirming its existence through his work by providing a representation of reality that is more solid because it is more coherent, that is, because all its manifestations are in better agreement with one another. And we have seen throughout this work to what extent an examination of the relativistic conceptions confirms this viewpoint.

196. *Any General Scientific Theory Appears to be Philosophical*

This powerful motive existing in the scientist's mind, although he is unaware of its constant influence, thus provides inspiration for any scientific theory that is at all general. As we pointed out in the case of relativism (§95), any scientific explanation is, and can only be, the small change of rationality, and by accumulating these explanations one reconstitutes the entire sum, whether one wants to or not. Consequently, any physical theory claiming to embrace a somewhat broad field is apt to appear 'philosophical'; moreover, as is demonstrated by the study of the evolution of such conceptions, the authors and supporters of the new concepts are frequently accused by their detractors of being 'metaphysicians' — this term, like the adjective 'philosophical' clearly assuming a pejorative sense in the mouths of scientists. But no matter how much science denies it, this 'philosophy' and this 'metaphysics' are an integral part of it; they are its own flesh and blood, and it would be futile to attempt to tear them out. Furthermore, if by chance science succeeded in this mad venture, one can declare in advance that humanity would immediately turn in disgust from such a science. Despite claims to the contrary, the goal of science is not merely to describe what happens; it seeks to understand why it happens, and it is only insofar as science accomplishes this that human intelligence is really interested in it.

197. *The Two Tendencies of Science*

Nevertheless, the attacks of which we have just spoken — particularly their frequency, their violence and the not too unfavorable way they are received by the opinion of the scientific milieu in general — show us that this desire not to overindulge in generalization and to stick as closely as possible to the observed facts also constitutes a natural tendency of the scientific mind. Although there is no science without reasoning and no reasoning without generalization, what truly characterizes the modern conception of science, as we have said, is the process of accepting the data as nature presents them to us, either spontaneously or as solicited by us, and studying them — in other words, understanding reality though observation and experiment. In this sense, Bacon and Comte were not altogether wrong in their protests and rendered a service to the physicist in warning him against his all too natural tendency toward excessive use of deduction. But of course they both went much too far, and the truth is that both tendencies are legitimate. True science embraces and contains them both; it needs them both. The only way it grows and progresses is through their interaction, or if one prefers, through the continual battle they wage within science itself.

198. *Why Chemistry Constitutes a Quasi-Independent Science*

A precise example will perhaps help clarify the meaning of this claim and show how well it corresponds to the real rhythm of the evolution of knowledge. We take it from the field of chemistry, whose origin we shall now consider from an altogether general point of view.

First of all, what is chemistry, strictly speaking, and how did it come to seem necessary to make a quasi-independent science out of this group of phenomena that obviously constitutes only a particular area of physics?

199. *Qualitatively Diverse Substances*

We said in Chapter 1 (§7) that the most direct and complete explanation of reality is the denial of diversity in both time and space. But we also granted that this approach, which is that of Parmenides, would reduce all science to absurdity. To avoid this, common sense creates the world of objects as a substratum, an explanation of changing sensation. Then science, in an attempt to make this explanation less inconsistent, is led first to consider change in time. It thus discovers that alongside the more or less accidental properties of matter, which appear and disappear easily, there are other properties that seem much more closely tied to the nature of the body, which disappear with more difficulty and reappear easily once they have disappeared. Now, as we know, science attempts to eliminate everything qualitative in its explanations and replace it by the quantitative. But in the case of chemistry, the qualitative reveals itself to be so essential, so deeply anchored in what to our mind must constitute the very being of the substance, that this substitution immediately seems much more difficult than in other cases. And it is this feeling, the necessity of assuming at least provisionally that there are qualitatively diverse substances, that characterizes the phenomena we class as chemical and leads us to separate them from the rest of physical phenomena. This is quite clearly demonstrated by the fact that all observations concerning the diverse chemical substances where their diversity (or specificity) is not of prime importance and where, consequently, these substances seem to be connected, are classed in a sort of intermediate science called *physical chemistry*.

200. *The Element as a Bearer of Qualities*

Thus we observe that qualitative concepts predominate in chemistry. At the beginning, that is, at the time this science is emerging from the semimagical prescriptions of its infancy, the chemical element is above all a bearer of qualities. At first the chemist considers only these qualities, completely neglecting anything having to do with quantity, or at least treating it as altogether secondary. The alchemists very logically use all sorts of procedures in an attempt to modify the qualities of bodies that superficially do not seem particularly distinctive, for example to transform an ordinary metal such as iron or copper into silver or gold; these attempts are exactly what is most characteristic of this ancient phase of the science of chemistry. But even after the futility of these would-be transmutations is recognized, the same spirit persists. Phlogiston is still only a quality-bearing element.

201. *The Element in Anti-Phlogiston Theory*

Lavoisier's reform, of course, chased this phantom from the domain of chemical theory. Must we conclude that this theory, as has sometimes been suggested, is henceforth freed of everything referring to quality? To see that this is not so, one need only make the simple observation that the first thing a chemist does when he wants to know the composition of a substance is to undertake an analysis that he himself calls qualitative. And a mere glance at a chemistry text is sufficient to convince us that this is not an empty term, that the component sought by qualitative analysis is really something assumed to be endowed with multiple properties: chromium is essentially different from lead, as iodine is from chlorine, and the very names of the last two elements are borrowed from the qualities attributed to them, whereas the case of chromium (which in its metallic form superficially resembles iron, the name referring to the vivid colors of its salts) indicates that the theory assumes a strict relationship between the properties of compounds and the essential nature of their elements. On the other hand, the persistence of the elements — the fact that they are conserved as such, with the qualities that are attributed to them and that are presumed to be simply masked or latent in the compounds — was certainly one of the essential foundations of nineteenth century chemistry. This affirmation was even an integral part of the law of the conservation of matter as it was generally understood (cf. IR 257 ff.; Eng. 236 ff.).

Moreover, this phase in the history of chemical concepts shows to what extent explanation is facilitated by the persistence of quality, as is consistent with what we explained in §199. Let us suppose that we want to account for the particular properties of sulphuric acid: we shall have incomparably less difficulty if we start from the elements sulphur, oxygen and hydrogen, each possessing well-defined properties, than if we were to reduce the whole to groupings of uniform atoms, let alone electrons. It is even clear that explanation becomes more difficult as the number of qualitative elements diminishes: it will be more difficult if we allow that sulphur is (as was sometimes assumed) a sort of polymer of oxygen than if we consider the properties of each element as simply given.

202. *Lavoisier, Prout and Mendeleev*

However, the above-mentioned assumptions, which constitute an obvious departure from the affirmation that the elements are essentially diverse, and which were already familiar to chemists of the preceding generations, clearly show us that even in this field the triumph of qualitative concepts was in reality much less complete than it first appeared. Granted that the whole of chemical theory, with the infinite multiplicity of its reactions, originated in the idea of the qualitative element, beneath this powerful and apparently unbroken wave there appeared a sort of undercurrent, strangely persistent in spite of very little encouragement from experimental results, or rather despite the fact that these results, and in particular anything apt to suggest the possibility of transforming one element into another, were constantly contradicted by experience. Already for Lavoisier there is a difference of rank among the elements; indeed, he seems to assume that certain elements, such as oxygen, nitrogen and hydrogen, are simpler than the others, which are consequently supposed to be derived from them. Some thirty [sic; in fact, not quite twenty] years after the death of the founder of modern chemistry, Prout set forth his famous theory: starting from the sound observation that if the atomic weight of hydrogen is adopted as the unit for all atomic weights, a great many other atomic weights seem to approach whole numbers, he asserted that hydrogen was to be considered the fundamental substance, all the other elements being only its compounds in varying degrees. This bold hypothesis was contradicted by well-established evidence; nevertheless it achieved immediate success and, though violently opposed, never completely disappeared from science. Almost a century after its inception it had the extraordinary fate of being not only formally revived but also generally accepted, since the apparent contradictions it had encountered somehow disappeared as a result of discoveries undreamed of by either Prout or his adversaries. On the one hand, Langevin showed that the loss of atomic energy resulting from the formation of the more complex atoms provides a plausible explanation of the small discrepancies that were observed,[3] and, on the other hand, Soddy's *isotopes* account for larger discrepancies. But much earlier than this very recent development, at a time when the persistence of the elements certainly seemed to be one of the essential foundations of chemistry, Mendeleev conceived his 'periodic table', which, by submitting all the elements to a rational classification based on their atomic weights, thereby implicitly affirmed that they could not be considered unrelated substances. "The periodic system of the elements," says Planck, "seemed to indicate clearly that there was only one ultimate form of matter" (PRB 42; Eng. 29). Sir Ernest Rutherford, in his recent presidential address before the British Association, also argued that Mendeleev's theory was "only explicable if atoms were similar structures in some way constructed of similar material."[4]

Although Mendeleev's theory at first met with strong resistance on the part of chemists, in the end it was almost universally accepted,[5] thereby eloquently attesting to the enduring strength of the undercurrent mentioned above. The strong growth of physical chemistry, and the more and more important place this intermediate science (§199) assumed in research as a whole, is evidence of the same sort.

203. *The Triumph of the Oneness of Matter*

Finally, as a result of a whole series of recent discoveries in the fields of both radioactivity and x-rays (cf. ES 1:224, 304), the venerable and fundamental concept of the oneness of matter has openly and apparently definitively triumphed. Of course, textbooks would still have us believe that there are qualitatively diverse chemical elements and that they persist throughout the operations we ordinarily perform in our laboratories. But at the same time they hasten to add that this diversity is not absolute, that in reality these alleged elements are not undecomposable. On the contrary, they are only compounds of a small number of ultimate components (generally thought to be positive and negative electrons), albeit compounds of a very different order from that of the substances we ordinarily synthesize or decompose in our laboratories.

204. *The Two Tendencies within Chemistry*

Thus chemistry was shaped by the constant battle — which is at the same time a collaboration — between these two conceptions: the one tending to affirm the diversity of substances, for which this diversity is essential and ultimate, and the other assuming that this diversity is only apparent and conceals a fundamental unity. And predictably, the latter tendency, growing out of a predisposition toward rationality, is the one more likely to inspire theorists, those who seek to reconcile the diverse data furnished by experience in order to create a coherent representation. On the other hand, those who observe and experiment are more likely to adhere to the qualitative conception.

This is in fact what happened in the case of Mendeleev's conception: we still remember clearly the time when laboratory chemists (including some of the most illustrious, such as Bunsen) never tired of heaping sarcasm on this 'chimera'. Going back a bit further, we can see an analogous situation in the period preceding Lavoisier. Physicists usually approach the phenomena we

now class as chemical from a completely general point of view; their commentaries, mechanistic by nature, clearly imply that quality has no objective
existence. Those who work more closely with things, however, and observe
the strong correlation between matter and some of its properties, necessarily
come to the conclusion that those properties are something substantial, that
they must persist through time and consequently produce no change except
through motion. In the quite conspicuous attitude of the physicist of that
time toward the chemist's findings there is, to be sure, something of the true
scientist's scorn for what he considered only an imperfect science, since the
chemistry of that time seemed an odd conglomeration, still imbued as it was
with the magic spirit of alchemy. However, there is also the fact of a strong
rational conviction, which does not intend to let itself be shaken by evidence
to the contrary, however warranted, because it *knows* that in the end these
discrepancies will necessarily be explained.

Finally, even today, if one looks closely, one can see traces of the same
kind of conflict. One day Georges Urbain remarked that "all a physicist
knows about chemistry today usually boils down to Mendeleev's periodic
table." This was obviously only a witticism on the part of the famous chemist,
but it nonetheless expresses, in an exaggerated form, an accurate and profound observation. For although the chemist fully accepts the oneness of
matter, as we pointed out above, he nevertheless knows that in the everyday
practice of his science he must continue to reckon with the distinct elements
of the chemistry of Lavoisier, endowed with many determinate properties,
only a very small number of which could really be deduced from the place
the element occupies in the periodic table, let alone from the hypothetical
structure of the atom. Will it ever be possible for this deduction to be complete? He hopes so, without always being completely convinced, feeling more
or less consciously that in the attempt to achieve this, science might well
run into difficulties that would force it to introduce into its explanations an
entirely unanticipated irrational element, which would appear unexpectedly
like Planck's quanta. At any rate, he sees here essentially a program, one to
which he is certainly willing to devote all his efforts, but whose full realization at the moment seems to him very remote. And then he is sometimes apt
to be surprised — or should we say annoyed — to see that the physicist,
accustomed as he is to looking at things from on high and concentrating on
properties common to all matter, speaks as if he believed the problem had
been solved, that is, as if he thought the behavior of chemical bodies in their
extremely complex relationships could actually be read from Mendeleev's
periodic table.

205. *The Analogy with the Evolution of Physics*

We beg the reader's indulgence for this long discussion of one particular aspect of the evolution of chemical theories, which he may see as a pure and simple digression, since the phenomena with which it deals are quite different from those treated by relativity theory. But, as we said in the preface, our main purpose here is to treat relativism in such a way as to expose the underlying thought processes. There can be no doubt that these thought processes are the same for all science, and it seemed to us that there was no better way to clarify the difficult question of the relationship between the Cartesian and Einsteinian deductions than to seek an analogy in another branch of science.

206. *Descartes Neglects the Specificity of Phenomena*

It would be quite unjustified to maintain that the way Descartes deals with optical, caloric and other such phenomena is completely identical with the way physicists of his time (himself included) treated chemical phenomena. Nevertheless, one can recognize something of the same spirit in both cases. Indeed, in reading through *Le Monde* or the *Principles*, whose explications are so admirable in other respects, one cannot fail to be struck by how little importance their author seems to attribute to the *specificity* of the phenomena he treats. Of course, he tries to take this into account to some extent, for example by the superposition of his different forms of elementary matter, but from our present-day point of view this is only a beginning. It is certainly no exaggeration to say that Descartes does all he can to avoid seeing this specificity, that in some sense he deliberately closes his eyes to it. And this is obviously what finally allows him to carry out his construction of reality, to mathematicize reality by absorbing it into space. We consider this a very important point, which deserves to be studied in more depth.

Borel, in putting forth the principles of Einsteinian mechanics, observes that "we may look upon the centrifugal force manifesting itself at the equator of the earth as a property of space and time; for the rotation of the earth, which is a physical phenomenon, can be expressed mathematically by means of formulas wherein space and time play a part" (ET 42; Eng. 35). This is absolutely correct and explains how Descartes was able to reach his goal without making use of any of the resources available to Einstein through higher mathematical analysis. Convinced as he was that everything in nature is motion, all he had to do was neglect the specificity of the different motions that must be assumed for a real explanation of the physical, or rather declare, by a sort of implicit postulate, that this specificity did not exist, because it *could not* exist. This is why his works, in spite of the remarkable discoveries they contain, particularly in the field of optics, nevertheless give us the impression of moving in a sort of empyrean, far removed from phenomena, at the very point where he claims to treat immediately observable physical phenomena.

207. *Experience Imposes this Specificity*

Why, after completely dominating European science for more than a generation, thanks to the admirable structure imparted to it by the genius of its author, did this system have to be just as completely abandoned so soon thereafter, and why was it subsequently almost forgotten? It can only be because the specificity neglected by Descartes could not be ignored. For experience, which, as we are well aware, played an altogether subordinate role for him, was soon brought to the forefront. The Baconian doctrine was instrumental in this change. To be sure, Bacon's influence on the development of science has been greatly exaggerated, for science, despite what has been claimed, is not Baconian; on the contrary, the programs of science are incomparably closer to those of Descartes than to the tables drawn up by the author of the *Novum Organum*. Nevertheless, it is incontestable that Bacon did influence the development of physics on this particular point, in reacting against the omnipotence of deduction (§197). And so, just as practicing chemists had to abandon the quite logical idea of the oneness of matter and adapt themselves to the assumption that there are diverse elementary materials with no link between them, those who observed optical, caloric and other such phenomena quickly came to understand that each class of these phenomena had characteristic features that precluded reducing them to Descartes's general system.

But where the reaction against Cartesian panmathematicism triumphed completely was in the Newtonian demonstration that even purely mechanical motion, if gravitation was not excluded, could be understood only by assuming the existence of something that was not purely spatial in nature, namely, a *force*.

208. *The Role of Newtonian Gravitation*

This phase in the evolution of scientific concepts is particularly instructive, because it shows us with great clarity what is necessary if the physical is finally to be reduced to the spatial. We realized in Chapter 14 that in order to be used in physics, mathematical deduction must have interpretation as an indispensable corollary, interpretation that transforms abstract magnitude into a physical *coefficient*. It is clear that in the current practice of science there is a vast range of interpretation. When a chemist writes Na + Cl = NaCl and adds grams of sodium and chlorine to deduce the weight of the resultant salt, he is not at all disturbed by the fact that he is dealing with three substances whose respective properties are completely different, nor that the numbers he treats as if they were of the same kind each has an appropriate qualifier. Similarly, in his calculations the physicist will substitute electrical energy for purely mechanical energy without the slightest qualm, even though he is completely convinced that they are quite different things. But when we are concerned not with partial, sporadic, limited deductions, but with a deduction embracing the whole of physical phenomena — in short, a universal deduction — the situation is no longer the same. For, as we know, such a deduction, if it takes place in the domain of science, can only be spatial. Consequently, it is clear that what interpretation adds to abstract number in this case will also have to be of a spatial, or geometrical, nature. If it is not, it will be impossible to fit it into such a deduction; it will be an alien and hostile element leading to the destruction of the very foundations of the deduction. This is what happened with the Newtonian concept of force. It is certain that when we posit the existence of force in the Newtonian sense, we more or less consciously try to make this concept agree with the concept of space, and even to show that it is derived from the latter concept. Thus we habitually imagine the force as, in some sense, spreading out over successive spherical surfaces (almost as light does), and from this image we deduce the inverse square law. Similarly, we like to picture force as a straight line and a center of force as a point surrounded by a sort of gigantic spider web. But it is no less certain that these are flawed contrivances. We are completely incapable of imagining what could possibly cause a perpetual flux of this sort emerging from a mathematical point, which is to say, from *nothing*, and it seems extremely difficult to understand how spider webs of this sort would be able to move through space without getting in each other's way (cf. IR 73 ff.; Eng. 75 ff.). The truth is that the action of Newtonian force takes place, in Lotze's picturesque image, "behind the back" of space;[6] the concept is fundamentally *aspatial*, or rather spatial and aspatial at the same time, that is to say, self-contradictory, which explains why (as we said in §60) physicists have constantly tried to eliminate it from their science. And it also explains why this concept was able to serve as an engine of destruction against the

Cartesian deduction of space: as soon as the concept of force is established in physics, the Cartesian system appears to be dead and forgotten. And likewise, we now understand why Einstein, no longer considering the phenomena of gravitation to be the effects of a force, but rather to be pure motion, was able to reintegrate them into the framework of spatial deduction.

209. *From Descartes to Einstein*

Let us briefly sum up the changes that occurred between the concepts of Descartes and Einstein. The former paid too little attention to the specificity of gravitational phenomena, being content to assume they could be explained by actions that were purely mechanical, as he understood the term. Then, by concentrating on this specificity, Newton destroyed the Cartesian system. Einstein returned to it by showing that specificity was only apparent in this case, since in the final analysis gravitation and inertia can be identified.

210. *Evolution in Other Areas of Physics*

A completely analogous evolution took place in other fields. After Descartes, the different areas of physics at first seemed to develop more and more independently of one another, so much so that at the beginning of the nineteenth century Auguste Comte, quite faithfully reflecting the prevailing opinion of laboratory scientists, as he did on other occasions, could state as a matter of principle that there were "six irreducible branches" of physics, "perhaps seven." All attempts to establish relationships between what today appear to be only divergent forms of a unique entity − energy − he condemned as due to "the long preponderance of the old philosophical spirit." [7] Nevertheless, it must be observed that at the very moment the creator of positivism was fulminating his anathemas, the work of unification had already begun and was even well advanced in several fields, which largely accounts for the harshness of his criticism. This work has continued untiringly since that time, and we know that toward the end of the last century physicists had come to believe that all the phenomena of nature could be reduced by the electrical theory of matter to a single fundamental phenomenon, the electrical phenomenon. We have also explained that this unification is exactly what finally allows Einstein's continuators − who considered the electrical phenomenon, as their master had the gravitational phenomenon, to be a simple phenomenon of motion − to dissolve the whole of reality into purely geometrical concepts.

It can now be seen how much this evolution, which is analogous for gravitation and the other forces of nature, resembles the evolution we described in the case of chemical concepts. There, too, the concept of a rational synthesis was first affirmed because the specificity of phenomena was ignored, then abandoned when this specificity was taken seriously, and finally reestablished following a long and difficult labor.

211. *The Role of Experimental Observations*

Science seems in some sense to have changed direction, since, after first attempting to understand phenomena in their diversity, it then seeks to abolish this same diversity. If we now ask ourselves what made this turnaround possible, it is certain that we must first of all invoke experimental observations. Indeed, it is obvious that the concept of the oneness of matter was able to prevail in chemistry because multiple relationships were established between the properties of elements — relationships summarized in Mendeleev's 'periodic table' and confirmed by Moseley's discovery — and also because it was observed in the study of radioactive bodies that substances being transformed into other substances nevertheless behaved in all other respects like genuine chemical elements. Likewise, it is the fact that Hertz was able to demonstrate experimentally that electrical waves behaved exactly like light waves that finally led to the identification of light and electricity, and the discoveries of Sir J. J. Thomson and Kaufmann are what motivated physicists to assume that mass was only an electrical phenomenon. Finally, the relativistic theories themselves were created under the influence of Michelson's experiment, and even if one argues that, as we ourself have pointed out, this was only an occasional cause, so to speak, since these theories could have arisen out of the consideration of Maxwell's equations alone, it is no less true that the theories were essentially the translation of the entire body of observations carried out in the field of electricity.

212. *The General Concepts Preexisted*

All this is quite true, of course. But is it the whole truth? Certainly not, for the whole truth also contains the observation that the general concepts of which we have just spoken existed prior to their triumph in physics. We know that the oneness of matter is one of the oldest ideas to occur to man as he reflected upon nature. The unity of physical forces is clearly implied by the assumption that everything is only motion, and in this sense is stipulated by the fundamental postulate of mechanism. The resolution of matter into space was clearly formulated by Descartes. Finally, the notion of hyperspace was perfectly familiar to geometers well before it was introduced into physics by Einstein and Minkowski. Of course these were ideas the physicist was inclined to consider purely philosophical or mathematical — one might as well say, *chimerical*, for they come to more or less the same thing for the laboratory scientist. Chimeras they may be, but it cannot be denied that they were real, live chimeras. A closer look will be sufficient to convince us that their life (like all life, it must be added) consisted of action, and it is through this action that the physical concepts we have just discussed came into being. Moreover, we have observed in Chapter 19 (§167) that the physicist, in response to tendencies of which he himself is not fully conscious, is perfectly capable of going beyond the conclusions that would seem to follow strictly from his observations. Now it is obviously such a tendency that we see here. Thus, although experience has certainly played a very large role in establishing the oneness of matter, it is not the most important factor. In this case experience was not free; it served reason and obeyed reason's aprioristic propensities.

213. *The Agreement between Reason and Nature*

This is a very general statement and one that deserves to be further elaborated, because, on the one hand, this state of affairs strongly influences the application of the method we constantly use, making it more difficult, and, on the other hand, it is likely to give rise to a misunderstanding of the results we achieve, by changing their meaning.

It might seem superfluous to dwell on the role of experimental discovery in the evolution of scientific concepts, given that it is clear — since we are concerned with science in the modern sense, which seeks to fathom nature by means of experiments — that the influence of the experimental is always implied. Moreover, we have had occasion throughout this work to stress the influence of experimental observations in the progress of explanatory science (cf. §167 in particular). We nevertheless believe it useful to state explicitly once again that the data of experience as presented by perception (more or less modified by the prevailing theory, if need be) serve as a basis — the raw material — for the work of reason, and that the principles that in our opinion guide this work then enter into play. For it is clear that they can manifest their influence only insofar as the data allow. These principles in fact appear as tendencies. Our reason attempts to apply them to — or, if need be, to impose them on — the reality it is in the process of understanding. In so doing, it does violence to this reality, but only to a certain extent, because the imposition of these principles would obviously be impossible if reality did not lend itself to it.

This (partial) agreement between reason and nature is just what we established in Chapter 4 of *De l'explication* ([ES] p. 111 in particular).

214. *The Misunderstanding on the Part of Positivism*

It can immediately be seen why, on this account, the principles directing scientific reason become harder to discern. Since they can actually prevail — find a clear and definite expression in the dominant scientific conceptions — only at a moment in the history of science when observation lends them its support and thus permits them to make their influence felt, does this not give rise to the temptation to disregard their influence entirely and attribute the creation of the theory to the effect of the experimental factor alone? Positivism can be said to be completely based on misunderstandings of this sort, given that it considers generalized experience to be the only legitimate content of science. Indeed, the powerful influence of this doctrine largely explains the widespread refusal to acknowledge the role of the *a priori* in scientific theories, even where it is most obvious. For example, we have the declaration that the conservation of matter is a fact discovered one fine day by Lavoisier, that the conservation of energy is only the generalization of an experiment, and that the 'scientific' conception of the atom, such as grew out of the observations of Brownian motion, has absolutely nothing in common with the 'wanderings' spoken of by Democritus and Lucretius.[8]

215. *The Utility of the History of Science*

Though we have pointed it out many times, it bears repeating that the best way to avoid these errors is by studying the history of scientific conceptions in as much detail as possible. One is most likely to understand the true nature of these conceptions by seeking to find out whether, and to what extent, they were able to be anticipated before they actually prevailed, on what grounds their proponents based their demonstrations, and what arms their adversaries used against these innovators.

216. *The* a priori *Factor in the Creation of Relativity Theory*

It can be seen, furthermore, that the influence of the *a priori* factor is perhaps even harder to ignore in the creation of relativity theory than in other cases.

Although it can conceivably be claimed that special relativity grew out of the experimental observations we have mentioned, this seems next to impossible in the case of general relativity. One cannot really point to the anomaly of Mercury, the only fact then known that was explained by the general theory and not already accounted for by the special theory. As we know, the other experimental facts, rather than being explained, were *anticipated* by Einstein; they are certainly brilliant confirmations of his theory, but could not have influenced the emergence of general relativity. Indeed, not only had the anomaly of Mercury been known for a very long time, but Einstein never claimed — and it seems highly unlikely in any case — that it was the possibility of explaining this phenomenon that gave him the idea of extending his earlier theory in the way he did. On the contrary, what directed him was obviously the desire to assimilate gravitation to inertia, and the resolution of the problem of the planet's motion occurred simply as a kind of by-product of this theoretical consideration. Thus, aprioristic tendencies did indeed provide the principal driving force in this case.

217. *The Aprioristic Tendency and its Concrete Realization in Science*

We now see clearly why the Einsteinian conception bears so strong a resemblance to Descartes's and why the relativist is nevertheless likely to seem shocked by this comparison, as if it somehow tended to debase the modern theory. The conceptions did indeed arise in response to the same tendency, but they represent very different stages in the influence of this tendency on physical science, different stages in its *concrete physical realization*, if we may use this term. The tendency first appears in the sphere of Parmenides; it is there that it is most absolute and most logical, but also furthest from the facts, which it frankly ignores. Since, as we have said, it makes all science absurd, the Parmenidean position disappears immediately, although not before giving rise to mechanism, which will henceforth be the dominant theory in science. Mechanism, after all, is only a sort of mitigated Eleaticism, the result of a fundamental compromise attempting to 'save' the reality of change by multiplying Parmenides' unique being. But then, with Descartes, the more rigorous form reappears: the reduction of everything physical to space. Unlike Parmenides' concept, this is a genuine physics, and an admirable one, which really explains a great number of very diverse phenomena. Nevertheless, this science does not take sufficient account of the diversity presented by the physical world, and that, as we have seen, is what destroys it. Finally, relativism appears, claiming to explain everything, the immensely large as well as the infinitely small; it claims to preserve the infinite specificity of the real phenomenon and nevertheless to reduce all of reality to space, to geometry — albeit an extremely complex geometry of which Descartes had no idea.

Is this the definitive realization? Who can say? The relativists certainly seem to be strongly inclined to think so. But it is easy to find parallels to this conviction in other epochs and concerning very different theories (cf. §133).

218. *The Tendency as Creator of Illusions: Transmutation*

We can easily observe in particular cases how readily the aprioristic tendency creates illusions, precisely because of its intrinsic strength. Let us limit ourself to a single example, again chosen from the field of chemistry. It is undoubtedly the belief in the oneness of matter that was responsible for the belief in the transmutation of metals universally held for so many centuries. People were firmly convinced that this assumption was confirmed by experiment. Now this was a great error, of course. It is true that we might have our doubts if we refer to certain popular accounts of recent discoveries, in which these discoveries are too often represented as implying a straightforward confirmation of the alchemists' theories. Such a representation is accurate in a sense, if one limits oneself to the general concept on which alchemy was founded. But when it comes to the possibility of transmutations as the alchemists conceived them, the situation is quite different. Intra-atomic energy is so much greater than the energy furnished by the sources currently at our disposal, and the amount of energy absorbed or emitted in the synthesis or decomposition of even a relatively minimal weight of elementary matter would be so formidable that it has rightly been said that if we succeeded in creating gold out of another element, the energy created or absorbed in the process would cost immeasurably more than the metal produced. Thus, far from confirming the alchemists' claims, this concept is a direct proof that they must have been mistaken and that there *could not* have been any transmutation during the operations they carried out using only procedures analogous to those of our chemical laboratories (but much less powerful, of course, than those to which we are now accustomed). One might even say that this is the first real demonstration that alchemistic transmutation is impossible. This impossibility, by the very fact that it involves an apodictic negation, is obviously very difficult to prove in any other way.[9] This observation serves well to bring home to us how profound the distinction sometimes is between diverse forms that one and the same fundamental concept is able to assume in science at different times.

219. *The Realization of the Idea for Plato and Hegel*

But the question of the scientific future of relativity, as we noted in our preface, is not, strictly speaking, our concern. What we must note, rather, is the constant realization of the *Idea*, in the Platonic sense of the term. Despite incessant contradictions inflicted on it by reality, the Idea tends to impose itself upon our conception of reality — to force reality to enter into the mold of the *Same* (which was the task Plato attributed to his demiurge), however unsuitable the mold might be to receive this reality — and to some extent it succeeds in this incredible undertaking. This process, moreover, resembles the one whose image is presented to us by human history, according to Hegel, since this history was supposed to consist essentially of the translation of ideas into concrete events by men and peoples entrusted, unbeknownst to them, with accomplishing this realization.

220. *The Disappearance of the Idea*

Thanks to the greater precision of scientific concepts and the greater understanding of their inner nature provided by the evolution of these concepts, we can now more clearly identify what became of this Idea when it was no longer manifest — or to be more specific, where it took refuge when it disappeared from science. It was lodged in the mind of the scientist. The physicist of the first half of the nineteenth century seemed to be convinced of the essential diversity of the forces of nature, just as the chemist was convinced of the diversity of his elements. But deep down inside them both, without their being aware of it, there certainly remained the desire to assert just the opposite. This is proved by the whole of the evolution that followed: the enthusiasm with which they sought demonstrations of the unity of forces or the oneness of matter, the eagerness with which they welcomed the affirmation of this unity, and the triumph of these concepts in spite of proofs that were often far from solid. There is no doubt that Einstein's cause is well served by this same state of mind.

NOTES

[1] Leibniz quite aptly characterizes this dominant trait of Descartes's effort when he says that Descartes had aspired "to the solution of the most serious questions in a kind of leap" (*De primæ philosophiæ emendatione, Opera*, ed. Erdmann, Berlin, 1840, p. 122 [*Philosophical Papers and Letters*, ed. and trans. Leroy E. Loemker (Chicago: University of Chicago Press, 1956), p. 708]).

[2] To understand the extent to which Cartesian concepts dominated physics at the time Newton appeared, one need only glance at the textbooks in use at the time; [Jacques] Rohault's text [*Traité de physique* (Paris, 1671)], which was accepted as the authority in all the countries of Europe for more than half a century, was entirely permeated with the Cartesian spirit (cf. Kurd Lasswitz, *Geschichte der Atomistik vom Mittelalter bis Newton*, Hamburg, 1890, 2:410).

[3] Paul Langevin, 'L'inertie de l'énergie et ses conséquences,' *Journal de physique théorique et appliquée* (5th series) 3 (1913) 584.

[4] Sir Ernest Rutherford, 'The Electrical Structure of Matter,' *The Times*, 13 Sept. 1923, p. 16, col. 1.

[5] The recent discoveries which, as we are aware, so clearly tend to confirm the concept of the unity of matter, nevertheless seem to some extent to shake the foundations on which Mendeleev built his theory, since the existence of isotopes poses problems for a classification based essentially on atomic weights. Indeed, as Alfred Berthoud says so well, the affirmation that "there exists a relation between the properties of an element and its atomic weight," which after Mendeleev seemed to be "an undisputable and definitively established principle," is now found to be "lacking in two respects," for "on the one hand we see the weight of the atom vary without its properties being affected," and "on the other hand we encounter elements that have the same atomic weight and different properties" (*La constitution des atomes*, Paris, 1922, p. 31).

[6] [Hermann Lotze, *Gründzuge der Naturphilosophie*, 2nd ed. (Leipzig, 1889), p. 26 (quoted in Milič Čapek, *Philosophical Impact of Contemporary Physics*, New York: van Nostrand, 1961, p. 95).]

[7] Auguste Comte, CPP 3:152, and *Système de politique positive* (Paris, 1851), 1:528.

[8] It would be superfluous to cite references in the case of Lavoisier; popular accounts often go so far as to attribute to him the maxim "nothing is created, nothing is lost." For the law of the conservation of energy, cf., for example, how it is introduced by Henri Poincaré, *Thermodynamique* (Paris, 1892), p. 65, and *La science et l'hypothèse* (Paris: Flammarion, s.d.), p. 157 [*Science and Hypothesis*, trans. W. J. Greenstreet (New York: Dover, 1952), p. 130]; by Gabriel Lippmann, *Cours de thermodynamique* (Paris, 1889), p. 11 ff.; and by Max Planck, *Das Prinzip der Erhaltung der Energie* (Leipzig, 1908), pp. 30, 41, 116, 105, 151. Finally, concerning atomism, cf. Marian Smoluchowski, 'Anzahl und Grösse der Moleküle und Atome,' *Scientia* 13 (1913) 27. For the psychological justification of this attitude, however, cf. below, Ch. 24, §270.

[9] From the beginning to the end of his address before the 1923 meeting of the British Association, Sir Ernest Rutherford treats the oneness of matter as an accepted and henceforth unshakable truth. Nevertheless, he arrives at the conclusion that it is quite possible that "uranium and thorium, represent the sole survivals in the earth today of types of elements that were common in the long distant ages, when the atoms now composing the earth were in course of formation." According to this hypothesis, "the presence of a store of energy ready for release is not a property of all atoms, but only of a special class of atoms like the radioactive atoms which have not yet reached the final state for equilibrium" ('Electrical Structure,' p. 16, col. 4).

CHAPTER 23

THE EVOLUTION OF REASON

221. *The Evolution of Reason for Hegel*

In an earlier work we showed how Hegel, in an attempt to explain the notion of becoming, to make it rational — or at least reasonable — had introduced the notion of a *concrete* reason: a reason of a kind hitherto unknown to logic. He had tried to justify this extremely bold innovation by arguing that the nature of human reason must have changed over the centuries through the very use humanity had made of this instrument. We endeavored to show that the claim of the German metaphysician was not tenable, that in fact everyone since the beginning of thought had used one and the same logic, and that the phenomenon of becoming in particular had remained just as opaque to reason after Hegel as it had been before this alleged institution of the concept of the 'reasonable'. At the same time, however, we admitted the theoretical possibility of some such change in the nature of reason, anticipating that if it should occur, it would be possible to discover it and determine its exact significance through an examination of the reasoning processes employed by science (cf. ES 2:379 ff.).

222. *Its Evolution in Relativism*

The considerations we have set forth here seem to suggest that with the advent of the theory of relativity we are really witnessing an evolution of this sort. For the relativist claims — in good faith, as we have said — to understand a significant part of becoming. And it can be claimed that he understands it because the instrument he uses to effect this comprehension, his reason, is no longer altogether the same as the reason used previously. Certainly the fundamental principles have remained essentially immutable. To understand how little inclined the new theory is to adopt Hegel's views, by assuming from the beginning that change in time is in conformity with reason, one need only observe that the relativists themselves often insist vigorously that only the permanent seems comprehensible.[1] And where it is impossible to establish identity in time, relativism captures the changing, just as one had always done, by resorting to spatial devices. But as a result of the modification of the concept of space itself, other possible means of explanation have come to be added to the traditional ones: the 'curvatures' or 'puckers' constitute resources that spatialization had not heretofore used.

This evolution obviously took place under conditions quite other than those Hegel saw as a context for the reform he imagined he could realize. The relativists have been careful not to proclaim the dawn of a new reason; on the contrary, they have done everything they could to maintain the belief that they reasoned just as one had always done, that they were merely applying (in Eddington's words) the "approved methods" of physics. But this does not prevent their adversaries from reproaching them for having recourse to illegitimate modes of explanation. One need only examine the arguments on both sides to see that in the final analysis the disagreement turns not on any one explanation in particular, but on the general fact that a mental construct one side considers to be a valid explanation appears to the other side to be absurd, or at least chimerical. It is therefore the very essence of reason as an explanatory faculty that is at stake.

223. *Duhem's Objections*

In speaking of the resistance met by the relativistic conceptions, we have dwelt especially on the opposition of thinkers unaccustomed to mathematical deduction. But this is certainly not the only kind of opposition to be found. On the contrary, an appreciable number of scientists also refuse to accept these theories, and we are now going to examine their arguments a little more closely (omitting those already discussed at the beginning of Chapter 8 who limit their criticism to the narrowness of the theory's experimental bases).

It is Duhem who seems to have expressed these opinions most clearly. Basing his argument on the fact that "one cannot in ordinary language and without recourse to algebraic formulae give a correct statement" of the theory (which is completely accurate, as we have seen), he observes that it "upsets all common sense intuitions" and "goes its way, proud of its algebraic rigor, looking with scorn at the common sense with which all men are endowed," whereas "in order for science to be true, it is not enough for it to be rigorous, it must start from common sense and end with common sense."[2]

224. *Science does not Return to Common Sense*

In our recent book (ES 2:182 ff.) we examined this point of view and came to see how difficult it is to sustain. Of course, it cannot be denied that science starts out with common sense perceptions and that the scientist begins by believing absolutely in the reality of the world of objects presented to him by this 'ontology'. It is equally obvious that the physicists' most extravagant hypotheses can have no other purpose than to explain the observations we make, whether directly by our sense organs or by these organs as they have been refined, reinforced, and in some sense extended by the use of instruments.[3] But it is not in this sense that Duhem understood the return to common sense. What he meant to affirm was that the theory, as it developed, had to end up with representations of reality that would be accepted without difficulty by common sense. In short, these representations were not to be too far removed from those offered by naive perception, and in any case they were not to violate any of the norms common sense recognizes as essential to its objective world.

Even a cursory glance at science and its evolution will show that the least contested theories, those most generally accepted by the physicists of a whole era, in no way correspond to this pattern. Let us consider Fresnel's hypothesis that light consists of the mechanical vibrations of a universal medium, the ether. This medium was supposed to be both infinitely subtle — since it offered absolutely no resistance to the motion of material bodies — and extremely rigid (even more so, strictly speaking, than steel or diamonds) since it had to vibrate transversally, as only a solid body could do. Was it not a denial of the most unassailable certainties of common sense to claim that one and the same substance could combine such contradictory properties, not to mention that, at bottom, the observation providing the firm basis for the theory, namely, that light added to light can produce darkness, is among the observations that have "appeared very paradoxical to common sense," as Borel so aptly put it (ET 149; Eng. 126)? Those who have written histories of optical theories have often expressed surprise at Newton's obvious bias for the corpuscular theory, or rather against the wave theory. Looking a little more closely, however, one realizes that he had very good reasons for this, the best of which was certainly the obvious fact that the hypothesis could not lead very far in the form Huygens had given it — that of longitudinal waves, similar to vibrations of the air. Before the hypothesis could be of real use to science, it was necessary for Young and Fresnel, by a device whose boldness is striking even today, to have recourse to the fundamentally absurd image just mentioned above.

The reader has also seen (§130) that even apart from relativity, the contemporary physicist is no less audacious in this respect than were his predecessors. Quantum theory, in particular, rests on foundations every bit as unacceptable to common sense (if not more so).

There is no doubt that although science starts from common sense, it by no means returns to it. On the contrary, one can easily see that it necessarily moves further and further away from common sense as it progresses. As a matter of fact, science modifies its concept of reality only grudgingly and as little as possible. This is why its early hypotheses bear so strong a resemblance to common sense objects, why atoms and molecules, as Bergson points out, seem at first to have been conceived for a sense of sight and a sense of touch altogether analogous to ours (§176). But as one follows the path of science, this 'anthropomorphism' (§22) is attenuated, and reality comes to bear less and less resemblance to the reality of our naive perception. One must not, therefore, be too surprised to find that once the relativist has scaled the heights of his hypothesis, he has a great deal of difficulty returning to the solid ground of this everyday reality (§91).

225. *Is Reason Immutable?*

But sometimes the objection we have just discussed is put more subtly. It is argued that it is not a question of the demands of common sense but of the demands of reason, which, for its part, can be supposed to be immutable. It is the principles of reason, it is claimed, that are contradicted by the funda-mental concepts of relativism, and it is certain that if this thesis could be established, the theories of Einstein and his continuators should immediately be removed from science.

226. *The Form of Reasoning*

In an earlier work we examined the question of how great a distinction can legitimately be made between the form of our reasoning and its content. To what extent can we separate the concept of the instrument by means of which we work, which is what we call *reason*, from that on which the instru-ment works (cf. ES 2:371 ff.)? We concluded that in the final analysis our thought is a whole, and that therefore these distinctions can only be made artificially. But we went on to say that the very fact of the existence of a science of logic nevertheless proves that our mind does make a distinction of this sort. For logic is obviously the art of reasoning about reason. It demon-strates the nature of the processes applied by reason; these processes therefore appear to be prior to the content to which they are applied.

One can, of course, consider that this priority is only relative and that in the long run *everything* that constitutes the intellect comes to it from outside, from sensation. In the words of the dictum so highly esteemed in the Middle Ages: *nihil est in intellectu quod non fuerit prius in sensu.*[4] But then, unless one denies any possibility of a logic, one is nevertheless forced to concede that what enters the intellect in this way constitutes two quite distinct classes of acquired knowledge, although this distinction becomes rather difficult to justify if the two classes have a common origin. For our present purposes the difference between this point of view and our own is largely verbal, which will enable us to simplify our presentation here. This opinion is that of Leibniz, who, as we know, added the words *nisi intellectus ipse* to the Latin dictum we have just cited. He thereby assumes that the human intellect itself has a content logically prior to all experience, constituting, according to Herbert Spencer's expression, "a notion given in the form of all experiences whatever they may be."

227. *Logic Proceeds* A Posteriori

It must be noted that we have no possible means of recognizing the totality of these forms *a priori*. Nothing is more certain (and we have stressed this throughout our books) than the fact that man does not experience himself reasoning, that he has no immediate consciousness of the path followed by his intellect. Only the results of this process, which is carried on unconsciously (almost like the workings of our digestion or our heart) reach our consciousness, in the form of decrees whose true causes are unknown to us. The same is true even for the simplest acts of our will, as Pascal so forcefully called to our attention.[5] Thus logic proceeds *a posteriori*: it is by examining particular examples of reasoning that it succeeds in knowing the nature of reason.

228. *The Process of Identification*

We might point out here that our own works provide some support for this claim. What we believe we have in fact established is that reason, when confronted with a phenomenon, behaves in an altogether predictable way: it attempts to explain the phenomenon first by making the consequent equivalent to the antecedent, and then by actually denying all diversity in space. Now this process, so well-defined and at the same time so important, so essential for understanding the nature of our reason, has been overlooked throughout the long centuries that reason, in all its forms — whether scientific, philosophical or common sense — was constantly applying it with unfailing vigor. Indeed, when we sought historical accounts of these conceptions, we could discover hardly any allusions to them. Such allusions as we did find among the thinkers of the past were generally rather vague and inconsequential; furthermore, they had almost no echo from their contemporaries or posterity. And we shall venture to add (at the risk of being accused of pleading our own cause) that the very reception accorded to our ideas constitutes a proof in the same sense. For although our works have not passed unnoticed, the thesis itself was apparently so far from obvious, and scientists (to stress only this side of the question) had so much difficulty identifying their own reasoning with the processes we had elucidated, that even today, sixteen years after the appearance of our first work, almost all science books are implicitly based on positivistic principles. Indeed they even present this theory explicitly (although it is quite inadequate to account for either the evolution of science in the past or for the present form of science), as if it were the only one that could possibly be considered.

229. *The Impossibility of* A Priori *Prediction*

Given that we must extract the principles of reason from particular examples of reasoning, including those of science, is there now any need to emphasize how difficult it would be to object to a process of reasoning current in science by an appeal to the principles of reason? The most we could say in such a case would be that, based on an analysis of the past, reason has until now proceeded in such and such a way. But *must* it unfailingly proceed in this way, and will these processes remain invariable in circumstances different from those of the past? Nothing about this can be known *a priori*; it is entirely *a posteriori*, by examining the reasoning of scientists and philosophers after the reform Hegel claimed to have introduced, that we demonstrated that — contrary to what he had asserted — he had not succeeded in making reason accept becoming. On the contrary, the attitude of reason toward becoming remained after him what it had always been.

In studying the functioning of reason, in trying to discriminate between that which belongs to reason in its own right and that upon which it works, one naturally tends to class under the latter heading everything that appears to have varied in our reasoning, reserving for the former everything that has remained constant (as we have pointed out, ES 2:374). But the question here is whether we can make such a distinction other than by analyzing particular examples of *actual* reasoning. Since we obviously cannot, any apodictic affirmation concerning the future behavior of reason therefore appears invalid. The best we can hope for is to arrive at probabilities that are larger or smaller depending on whether they are founded on a more or less comprehensive and exact analysis of the past behavior of reason.

230. *Are the Concepts of Time and Space Part of Reason?*

In the more particular case of the relativistic concepts, the opposing argument of which we have spoken obviously claims that our familiar concepts of time and space are part of the essence of reason that must not be disturbed. This approaches the position of Kant, for whom the forms intuition imposes on reality certainly did not seem to be variable — we are again up against that aspect of his doctrine, unfavorable to relativism, which we discussed in Chapter 18.

But was Kant completely correct? We believe, on the contrary, that careful consideration forces us to acknowledge that at least the concept of space, not only for the modern scientist but also for many generations of his predecessors, is not altogether the same as the one on which naive perception is based.

231. *Up and Down*

Let us begin by thinking about up and down. Can one seriously question that the primitive concept of space includes the fact that this double direction is essentially distinct from the others? To be convinced of this one need only examine the objections raised against Christopher Columbus. At that time, of course, the fact that the earth was spherical was universally accepted, and had been for at least twenty centuries, since it was already considered established in Plato's time, and the two cosmological systems taught during the Middle Ages — Aristotle's and Ptolemy's — both assumed it. However, this did not prevent people from arguing that as the explorer went west he would slide down a mountain of water and not be able to climb back up. Moreover, each of us can recognize the vestiges of an analogous feeling in his consciousness. For although we are firmly convinced of the sphericity of the globe we inhabit, we are nevertheless incapable of picturing the antipodes except with their heads down and their feet sticking up. Thus, for our immediate intuition, up and down still remain privileged and unchanging directions in an absolute space.

232. *Properties of the Vertical and Geometrical Properties*

It can, of course, be argued that, in spite of this belief, there was never, from the very beginning of science, any confusion between this property of space and those we qualify as geometrical. However little we know about the beginnings of Greek geometry, there can be no doubt that these beginnings go back much further than the doctrine that the earth is spherical, and certainly no geometer ever felt the need to prove that a proposition valid for a triangle drawn horizontally was still valid for the same triangle in a vertical plane. But this proves only one thing: that people had very early become accustomed to making a distinction between the property of space responsible for the existence of up and down, and all its other properties. The separation was extremely advantageous precisely because it permitted the establishment of geometry, and the brilliant success of this branch of knowledge tended, of course, to confirm the conviction that the distinction is absolutely essential. This conviction is so great that even though it is well-known what a dominant role Kant attributed to spatial intuition, he was nevertheless able to exclude anything having to do with up and down by a sort of preterition. But this does not prove that it was originally excluded. Indeed, the facts we have cited tend to establish that naive intuition does include the specificity of the vertical direction among the essential properties of space.

233. *The Evolution of Reason Brought About by Inertia*

From what we have shown in Chapter 4, it must also be assumed that the introduction of the principle of inertia involved an analogous evolution. This evolution, too, is masked by the fact that, when we speak of spatial intuition, we are inclined to make a distinction between the physical properties of space and its mathematical properties. But here the authority of Kant turns against him, so to speak, for as we showed in §36, Kant completely deduced Cartesian inertia from the intuitive concept of space. Without doubt a confirmed Kantian would respond that the master nevertheless in no way meant to allow that this concept had changed; on the contrary, he obviously assumed that spatial intuition had always been just as he defined it. But we have also seen that the attitude of all pre-Galilean science (up to and including More) protests against any such claim. Even today there are physicists like Eddington who are able to realize that their predecessors did not simply 'make a mistake' on this issue by paying insufficient attention to their own convictions, but that their feeling was perfectly legitimate. Here, as in the case of up and down, the concept of space has most certainly changed as human knowledge has progressed.

234. *Physical Space and Geometrical Space*

Certainly the two specific traits of which we have just spoken, the existence
of up and down and the behavior in space of a body set in motion, concerned
physical space. But can one claim on the basis of this distinction that only
what has to do with physical space is subject to change, while the foundations
of the geometrical concept of space (that is, everything Kant meant to
embrace in his intuition) must, on the contrary, remain eternally immutable?
We noted above how hard it seems to be, from the standpoint of intuition
itself, to separate the idea of up and down from the rest of the ideas constituting the notion of space. But there is also the fact, which is surely quite
remarkable if one will but take note of it, that the concept of hyperspace (as
we indicated in Chapter 20) was initially an entirely disinterested speculation,
without the least pretention to physical explanation. Thus it cannot be
denied that the concept arose out of geometry itself, that it is the very fruit
of this branch of knowledge, which, if one assumes the existence of spatial
intuition, is and can only be a systematization and development of this
intuitive concept of space.

235. *Proclus and Euclid*

The history of non-Euclidean geometry is extremely instructive on this point.
It is certain that the origin of the ideas that gave rise to this geometry goes
back much further than the geometers of the first half of the last century
usually cited as its authors (Lobachevsky, Bolyai, Riemann). Hints of it are to
be found even in Greek antiquity, since Proclus, as Weyl correctly points out
(Chapter 20, note 3) already expressed doubts about the theorem [sic] of
parallels. But did not Euclid himself somehow pave the way for this later
evolution by attributing a particular character to the proposition stipulating
that only one parallel to a given straight line can be drawn through an external
point? Did he not show that this notion could be distinguished from all the
others by which we express our concept of space? Borel commented with
some justification that having established this distinction was, "perhaps, the
most brilliant proof" of the "deductive[6] genius" of the great Greek geometer
(ET 51; Eng. 43), and there is no doubt that this was actually the starting
point for the whole evolution. At any rate, one can see the weakness of the
argument according to which Euclidean geometry is considered to be an
untouchable whole, given that Euclid himself already recognized the heterogeneity of the notions on which his geometry is based.

236. *Reason and Sensation*

Since it is impossible for us to determine *a priori* what in our reason must remain immutable, we are therefore forced to consider that reason itself is capable of evolving. It evolves by abandoning requirements that no doubt originally seemed to it to be essential, peremptory.

However, it must be understood that reason has no intention whatsoever of denying the existence of the elements it sets aside in this way; it merely pushes them back toward reality, that is, sensation. Thus, when the geometrician eliminates the distinction between the vertical direction and all other directions in his reasoning, he is by no means denying that our immediate intuition recognizes such a distinction. Instead it becomes for him a purely physical fact, due to our habits, the fact that we stand erect, the importance assumed by gravitational phenomena in general among the phenomena we observe, etc. − in other words, due to considerations that do not concern him because they are imposed on him from outside and are therefore considered alien to his deductions. By the same token, when the Cartesian affirmed that, contrary to what had previously been assumed, space was indifferent to uniform rectilinear motion, he was not contesting that immediate observation seemed, on the contrary, to show motion stopping of itself; rather he was explaining that this was an appearance due to friction and air resistance. These are purely external and accidental circumstances, while, on the other hand, the behavior of space with regard to motion is supposed to be a part of the essential nature of space itself. Finally, the relativist does not at all overlook the fact that our sensations, as they are spontaneously transformed into perceptions, place themselves in a three-dimensional space entirely free of all curvature; what he claims is that this very circumstance is what makes us believe that this concept of space is imposed on us by our reason. According to him, however, this is only an illusion, since our concept of space is actually broader and perfectly compatible with the existence of these curvatures that, on the other hand, a more rigorous examination of the phenomena requires us to assume. Once again, then, we find eliminated from reason those postulates that, when added to the general concept of space formulated by hypergeometry, transform it into Euclidean space. These postulates are transferred to sensation − to experience − so that this part of spatial intuition (as understood by Kant) is no longer anything more than a generalization from experience.

Since it is physical observations that lead us to the hypothesis of curvatures, one might be surprised that immediate perception provides us with no trace of this notion. Recalling Weyl's statement quoted in Chapter 6 (§67), but which he applied instead to the concept of a four-dimensional spatio-temporal universe, we might say that it is strange that three-dimensional geometry seems altogether obvious to us while higher geometries are so difficult. But all things considered, this observation is not much more paradoxical than the fact that perception strongly inculcates in us a belief in the existence of up and down and makes us believe in the nonpermanence of motion.

237. *The Conviction Created by Success*

Once reason has pushed these elements outside itself one after another and this operation has been sanctioned by the common consent of competent opinion, we obviously tend to claim that they were never really part of the instrument by which our intellect operated — namely, reason itself — and that they belonged instead to the content on which reason operated. To realize this, one need only consider Kant's attitude and note how little inclined he is to assume that spatial intuition could ever have included the affirmation that up and down exist or that space has any inhibiting effect on motion. But it is quite certain that we can arrive at this sort of conviction only after the fact. What then creates the conviction is obviously nothing other than success: by consenting to abandon some of the requirements of our reason — that is, we repeat, to modify our reason, the instrument by which we try to grasp reality and explain it — we succeed in grasping this reality better and explaining it further.

238. *Has there been Analogous Evolution Outside the Sciences?*

We can now see that all the modifications of reason discussed in this chapter deal exclusively with what could be called geometrical reason, that is, with the way our intellect conceives of space. It could of course be claimed that this results from the way we have approached the problem. Indeed, in our preceding book (cf. ES 2:379), we indicated that the best way for the philosopher to ascertain whether or not the processes of reason have changed is by examining the procedures used by science. This is obviously the method we have followed here. Now science attempts to understand reality by mathematicizing it; its ideal is a pangeometrism, and it is therefore only natural that our scrutiny of the evolution of science revealed only variations involving geometrical reason. But this by no means proves that a similar effort undertaken outside the sciences properly speaking (that is, the mathematical and physical sciences) would not lead us to recognize analogous changes in the apparatus of thought.

One's first response to this might be that, as a matter of fact, no modification of this sort has thus far been able to be identified – though not for the lack of trying. At the beginning of the chapter we recalled how Hegel had thought he could establish that an evolution had occurred in the way reason deals with the concept of becoming, and we said that we had been able to demonstrate by means of concrete examples that this claim could not be upheld. Although many very distinguished thinkers after Hegel were seduced by this idea of an evolution of reason, one would search in vain among the supporters of this concept for a demonstration of such a change, let alone an example that is in any way precise.

Will it be objected that this is due simply to the somewhat extraneous fact that since scientific reasoning is by its very nature more precise, any modifications liable to occur in it stand out better and are more easily exposed? In fact, would we not make quite similar observations if we were able to analyze extrascientific reasoning just as closely? Let us add immediately that as far as the inferences to be drawn from it are concerned, this point of view would differ from our position only formally, so to speak. Indeed, since the workings of reason intervene in all our thoughts, it is obviously in our best interest to claim that this process is unchanging if we want to understand anything about our forefathers, not to mention reach any understanding with our contemporaries. Only a peremptory proof would make us abandon this claim; therefore, if the modifications that have occurred are so subtle that analysis has been unable to uncover them up to now, we shall have to declare boldly that there have been none.

239. *The Objective and the Subjective*

But that is not all. This analysis would seem to suggest that the fact that only geometrical reason appears to have evolved with time might actually result from the very foundations on which our rational conception of reality is built. Indeed, we have just seen that the evolution of reason consists essentially in eliminating elements that were until that time supposed to be part of it and are thereafter recognized as coming to it from outside. Now this obviously implies that in the region where the transfer took place, it must have been particularly difficult to determine the boundaries between reason and reality – between what reality brings to the work of reason and what reason, for its part, adds to it. But this requires further explanation.

Given that philosophy and, after it, science — which is, from this point of view, only a particular form of philosophy — are essentially speculations about the nature of reality, it is clear that the question constantly being posed is actually the question of the distinction — in the raw material of spontaneous perception taken as a whole — between what was really the observed, the object, and what, on the other hand, was brought to it more or less unconsciously by the observer, the subject. When a scientist, after having *described* an observation, that is, noted what it is that perception in- itially brought him, then goes on to *examine* it, he is making this distinction; he is implicitly declaring that certain elements, though incontestably con- tained in his perception, did not belong to reality but to himself.

240. *The Give and Take between Reality and Sensation*

But it is important to point out that this give and take between reality and sensation occurs constantly. When Democritus declares that anything qualitative can only be opinion, since only quantitative properties can be attributes of reality, or when the modern scientist attributes our particular sensation of the color red to 'the specific energy of the nerve' (these are two fundamentally very similar forms of one and the same aprioristic conviction), what is thus subtracted from the object and added to the subject in no way increases the domain of reason. It would seem absurd to claim that reason required us to attribute the property we call the quality of red light to a thing which in itself we do not now even consider to be light, and which can, on the contrary, give rise to other sensations that are not even luminous.

241. *It Takes Place in the Spatial*

Let us now recall the observations we made in Chapter 3 (§23) concerning the obvious advantage of spatial over extramathematical deduction when it comes to establishing a theory of reality. We have seen that this advantage lies in the fact that the spatial exhibits an agreement between the mind and reality, a fact on which scientists and philosophers, according to their individual preferences, have frequently based their demonstrations either that an external world exists outside the self or, on the contrary, that this world depends on thought.

Consequently, this would seem to force the conclusion that the same fundamental fact also explains why it is possible to transfer this or that component from *a priori* reason to experience in the spatial realm. It is because the mind and reality seem to be in perfect agreement in the spatial that we find it difficult to determine what must be attributed to the one source and what to the other and that we can, in the final analysis, change this attribution by arguments that are more or less supported by experience.

242. The Compliance of Reason Remains Incomplete

Thus transfer is *possible* here, but it is certainly far from easy, for these elements of spatial intuition are obviously very solidly rooted in our minds (which is the very reason we can consider them to be inseparable from it). They are so solidly rooted that in this case the compliance of reason is not absolutely complete. As we have observed, even today, after so many centuries of mathematical education, man can still find deep within himself the ancient prejudice in favor of the privileged status of the vertical direction in absolute space. Similarly, it is certain that Cartesian inertia – the concept that space has no influence on a body in uniform rectilinear motion, so that once a mass is set in motion it must continue this motion eternally without purpose and without end – is also somehow distasteful to the mind. This is obviously why, as Eddington observed (§30), we find it more natural to assume that motion runs down in the course of time. Finally, in the case of the relativistic theories, it would be superfluous to dwell on the resistance they encounter. But even if these theories were to achieve the most complete triumph imaginable, and even if the scientific community (however broadly one might conceive it) had been accustomed for centuries to consider the whole body of physical phenomena in these terms, it would seem possible to predict without risk of error that when these scientists opened their eyes in the morning, they would still never perceive objects except in a three-dimensional Euclidean space, that is to say, one free of any curvature.

243. Why Reason Consents to the Required Sacrifice

Why does reason nevertheless agree to the very difficult task of changing its own nature? Why does it resign itself to a sacrifice it obviously considers extremely painful? There can be no doubt on this point: it is because this sacrifice seems to be more than offset by what reason takes to be a gain on another side; it is because it considers this sacrifice the inevitable price of a victory. Reason is by nature absolute; it does not allow anything at all in reality to escape it. Everything must be subject to it, everything must appear rational. This is the fundamental postulate of human knowledge and consequently serves as the basis for both science and philosophy. Both start from common sense, which is in fact only a sort of unconscious and rudimentary science and philosophy — reason's first attempt to grasp reality. But this system quickly appears inadequate to us, for although it explains some phenomena, others that seem just as important clearly escape it. Then, in order to include these phenomena, we abandon the primitive system, substituting other systems which, if they no longer have the advantage of arising in us spontaneously, like the one implied by immediate perception, have another advantage that seems to take precedence: they are more comprehensive; they extend the limits of the totality (§15) we embrace.

It is obviously this same tendency that requires us to modify the nature of reason itself. Reason resigns itself to the necessity of changing its own nature because it becomes convinced that by so doing it will be able to master a side of reality that hitherto eluded it.

244. *Reason Abandons Elements that were Part of It*

In so doing, reason enlarges its domain. One might say it enriches itself, if one considers only the extension of the limits of what it succeeds in comprehending. It must be understood, however, that insofar as the actual content of reason is concerned, the modification is really an impoverishment. For reason changes only by abandoning some of its postulates, that is, by cutting out something that previously seemed to be part of it. The concept of purely geometrical space is poorer than the one stipulating that up and down are distinct directions, just as the concept of Cartesian space is poorer than the Aristotelian concept, which implies an inhibiting effect on the motion of bodies. And, as to the non-Euclidean space of relativism, we know that it is reached by eliminating the postulate of parallels, in other words, by rejecting a property that was previously part of the concept of space.

245. *The Content of a Concept and its Extension*

This relationship between the area taken in by reason and reason's actual content is not at all surprising. It is analogous to the relationship we see in logic between the content [or intension] of a general concept and its extension — and for good reason, since we are obviously dealing with the same sort of thing. As we move from the concept of species to those of genus, class, etc., we gradually leave behind the distinctive peculiarities characterizing the narrower divisions, so that the broadest concept at the same time necessarily has the least content.

246. *The Modification of Reason for Hegel and for the Relativists*

It should be noted, moreover, that these last remarks, unlike those that preceded them, no longer refer to geometrical reason alone but to reason in general. One can see that this is true by noting that the modification of extramathematical reason that would have resulted if Hegel's project had succeeded would have taken place in the same direction as the evolution of geometrical reason. Hegel tried to persuade us that reason had modified its nature, that it immediately comprehended becoming or was at least capable of doing so, showing what an immense area reason could thereby add to that which had always been considered to belong to it, and how all of reality could consequently seem *rational*. For him, this brilliant conquest obviously more than counterbalanced the arduous constraint reason had to exercise on itself: the very extent and intensity of the effort required prove that he did not underestimate this side of the question. On the contrary, he essentially realized that it would take powerful motives to induce reason to accept this sacrifice.

At the same time it is clear that what Hegel was postulating constituted an impoverishment of the content of reason. Reason seeks identity and postulates its existence wherever sensation makes us perceive diversity. Now it is this postulate that Hegel's *concrete* reason was required to eliminate unequivocally from its essential nature.

247. *Our Attempted Reform*

It might seem in some respects that, by attributing to reason a requirement that had not commonly been considered characteristic of it until now, we have attempted a reform similar to the one undertaken by Hegel, but in the opposite direction, since the content of reason is thereby augmented. There is, however, the essential difference that we do not claim that a new form of reason is involved. We make just the opposite affirmation: that, insofar as its forms are concerned, reason has been identical to itself in all ages and in all circumstances. In other words, what we have sought to do is to enrich the theory of the forms of reason, but not to increase the actual number of these forms.

248. *The Utility of a Comparison with Hegel*

It can now be seen that in spite of all the differences separating the evolution

of reason implied by relativity theory from that contemplated by Hegel —
and these differences are altogether essential, as we have seen — there is
nevertheless a basis for comparison in the very fact of the evolution and
also in the fact that in both cases this evolution takes place in the same
direction. But here we must explain ourself. We related above how Hegel,
while claiming that the nature of explanatory reason could change, had
largely succeeded in distinguishing the true principles guiding scientific
reason. But of course he had not connected the two concepts, or rather, the
change he envisaged had no bearing on scientific reasoning strictly speaking
(or what we would today qualify as such). For the foundation of explanatory
science had seemed so paradoxical to him that he had come to the conclusion
that the only possibility was to reject this science altogether and seek true
explanations in a totally different form of knowledge. And is it not a real
irony of fate that the evolution of reason, which he seems to have been the
first to consider possible, came to be realized through a development of
mathematical theories, which he considered futile and would have liked to
cut out of physics? This must not deter us from admiring the depth and vigor
of thought he was able to display in a work that an initial error prevented
from exercising any influence whatsoever on the progress of either science
or epistemology. Insofar as the theory of relativity in particular is concerned,
it would certainly be excessive to consider Hegel a precursor, or even a simple
predecessor, for there is really too wide a gap between certain very essential
tendencies of the two theories. But at the same time this must not prevent us
from using the Hegelian theory for comparative purposes. We believe we have
demonstrated that such a comparison is eminently suited — precisely because
of the equally striking similarities and dissimilarities — for making clear the
thought processes used in each case.

NOTES

[1] "Mind exalts the permanent and ignores the transitory," Eddington declares (STG
198); cf. also similar passages, STG 141, 196.
[2] Pierre Duhem, *La science allemande* (Paris, 1915), pp. 135, 136, 143.
[3] We developed this point a bit further in our article on 'Hegel, Hamilton, Hamelin et
le concept de cause,' *Revue philosophique* 96 (July–Aug. 1923) 52 ff.
[4] It must be noted, however, that many medieval thinkers held a position similar
to that of Leibniz on this point. St. Thomas, for example, stressed the active role of
the intellect: *Sed quia phantasmata non sufficiunt immutare intellectum possibilem,
sed oportet quod fiant intelligibilia actu per intellectum agentem; non potest dici quod
sensibilis cognitio sit totalis et perfecta causa intellectualis cognitionis; sed magis quo-*

dammodo est materia causæ (*Summa Theologiæ*, Pt. 1, qu. 84, art. 6, *Opera Omnia*, Rome 1889, 5:324).

5 Pascal, *Pensées et opuscules*, ed. Brunschvicg (Paris, 1917), p. 457.

6 [Although "inductive" appears in the Rappoport and Dougall translation, the French text reads "déductif."]

CHAPTER 24

DOGMATISM AND SKEPTICISM IN SCIENCE

249. *The Perplexity of the Layman*

What in the preceding pages we have called simply reason is, needless to say, a particular kind of reason, namely, the reason of the scientist or, more precisely, the theoretical physicist. In the course of this work we have alluded several times to the fact that there is a considerable gap between this reason and that of the ordinary man, and even the man who is considered educated in the ordinary sense of the word, but is insufficiently grounded in the study of the sciences. We have stressed in particular that the imagination inexperienced in this way of thinking is unable to achieve any real understanding of the relativistic concepts.

But if this is the case, what attitude must the layman take toward relativity? It is useless, as we have said, to prescribe forbearance. What then is he to believe? In what sense and in what measure must he transform his everyday concept of reality which is as necessary to him as the air he breathes?

250. *Science is Both Dogmatic and Skeptical*

This problem, if one considers it in its most general form, is the question of what value should be attributed to scientific pronouncements, and it is only fair to acknowledge that the nonscientist has a right to be somewhat at a loss here, given that he is confronted by affirmations, or at least attitudes, that seem hard to reconcile. On the one hand, he sees scientists lay claim to a quasi-absolute authority for their judgments; not only do they pay very little attention to the most deeply rooted scruples of our common sense, but they scorn, or entirely ignore, any objections stemming from a point of view and a method different from their own. They do not hesitate to do this even when these objections are presented by men who in other respects command a considerable reputation and influence, as can be seen in the case of Goethe or Hegel. On the other hand, however, these decisions that seem to have been laid down as articles of an absolute faith obviously do not have any permanent status for the scientists themselves. For all sorts of reasons, and often even for reasons that seem to the layman to be incomprehensible or quite trivial

216

and, in any case, ridiculously inadequate as a basis for settling such weighty problems, scientists abandon without a qualm a system that a short time before seemed completely secure. And once this about-face has been accomplished, they treat as futile and entirely unworthy of notice any attempt to return to concepts that had for many years been taken to be infallible expressions of the truth, not only by science but by all of humanity, precisely because it obeyed the injunctions of science. To see the truth of this, one has only to put to a contemporary astronomer an assumption entailing Ptolemy's system, or to a chemist an explanation based on the existence of phlogiston. When such dogmatism and such skepticism are professed simultaneously they might seem to destroy each other. In any case, the two of them together are fairly likely to be disconcerting to the layman. Recent debates such as the one following Henri Poincaré's definition of scientific truths as simply *convenient* (cf. Chapter 2, note 9 above) offer ample evidence that it is not only the layman who is disconcerted.

251. *The Extravagance of the Relativistic Ideas*

The theory of relativity seems particularly well suited for elucidating this subject. First of all, since relativity has not yet acquired the full authority of an established theory, its principles will be harder to confuse with the principles spontaneously decreed by our reason. Secondly, and more importantly, it is characterized by what we might venture to call its extravagance, because of the divergence between the concepts it would impose on us and those we possessed until now. When we profess to believe in the existence of the planet Mars, it is absolutely certain that this object, in the form given it by astronomy, was not a part of our direct perception. What one sees if one looks at the sky is a reddish luminous point, and quite sustained observation is necessary to see that it constantly changes position with respect to the ensemble of other stars said to be fixed because they maintain the same positions in relation to one another. Beyond these simple details, everything we claim to know about Mars is scientific in origin, which is to say that it has come to us through observations made by specialists, especially with the aid of instruments — which, however, sometimes have a strange way of leading us astray, as for example in the case of the famous Martian 'canals' — or through conclusions drawn from these observations. Nevertheless, the picture the astronomer has formed by means of these observations and conclusions — which is that of a planet very similar to the earth with an atmosphere, seasons, water, etc. — is so much like what is familiar to us that it seems to take a place more or less spontaneously among the representations making up the whole of the common sense world (as we have noted in §16). It is certain that at least educated men, although most of them have never even seen the planet Mars through a telescope, are inclined to speak of it as if they had walked upon its surface. There is obviously no fear of this kind of confusion in the case of the relativistic concepts.

252. *The Conflicts between Science and Theology*

Earlier we used the term faith with respect to scientific theories. Indeed, the most superficial observation shows that there must be some analogy between the decrees promulgated by science and those belonging to faith strictly speaking; if need be, this could be proved by merely looking back at the often quite violent quarrels that have broken out in the course of history between proponents of these two ways of thinking. Such battles, of course, usually resulted from the purely extraneous circumstance that a scientific claim was in contradiction with the way things were expressed in Holy Writ. By looking more closely, however, one sees that there was also a deeper opposition: as Tycho Brahe, for example, rightly argued, the place our planet occupies in the Copernican system is difficult to reconcile with the ideas inspiring Biblical cosmography and cosmogony, according to which the earth was the principal object of divine activity and as such was contrasted (in the well-known verse from Genesis) with the heavens in their entirety.[1] On the other hand, these discussions have obviously abated for the most part, and a boundary has been drawn, after a fashion, between the two domains, completely closing them off from one another. But there is no doubt that this has been a rather difficult process, and although many men who are both sincere belivers and strongly devoted to science do not have the slightest qualms of conscience, for others a painful conflict persists. Further proof that the former state of mind has not entirely disappeared is to be found in the fact that very slight concessions (even merely apparent ones, as was the case with Henri Poincaré's statements mentioned above) are sufficient to give rise to renewed boundary disputes. Now, in order to be opposed to one another in this way, it is obvious that the two sorts of truths must belong, is some respects, to the same class.

253. *Science as a Substitute for Religion, According to Comte*

This was indeed the opinion of Auguste Comte, who, as we know, went almost too far in this direction. For him, as a matter of fact, in the social and political state instituted by the triumph of 'positive' thought, theological truths were to be entirely replaced by scientific truths, which would form the unshakable foundations of the new edifice. In order to accomplish this, however, scientific thought itself had to undergo a certain transformation; in particular it had to leave the somewhat anarchic state in which it seemed to be struggling. It had to *get organized*, as we would say today — we do not feel that it is a betrayal of Comte's position to use this term, despite the slightly ominous meaning it acquired as a result of certain demands preceding and following the 1914 war. What Comte actually envisaged was a true organization in the authoritarian sense of the term. The scientific beliefs of the public and, to an even greater extent, the research work of the scientists themselves were to be strictly regulated and supervised by a public authority composed of men accredited by the state and provided with all the sanctions of the secular arm. This regulation, of course, like all regulation everywhere, was to consist primarily of prohibitions, and Comte drew up a preliminary list of them: It is forbidden to carry out any investigations that are not 'positive', that is, all research must be limited to the search for laws; it is forbidden to attempt to understand problems that man clearly has no need to understand and that must therefore remain entirely opaque to him, such as, for example, the chemical constitution of the stars; it is forbidden to study phenomena that are too minute, in particular with the aid of the microscope, which is only an 'equivocal' means of investigation and which is given an 'exaggerated value'.[2]

254. *The Importance of Disinterested Research*

Attempts have sometimes been made to excuse Comte's attitude by arguing that even though we might be inclined today to condemn his methods, on the whole he merely meant to direct scientific activity toward goals he considered much more useful than others. Indeed, at first glance his bias seems to follow directly from the very foundations of the positivistic philosophy, which, as we know, follows Chancellor Bacon in defining science by its utility with respect to action. But we all feel that nature is a coherent whole in which all relationships are interdependent, and that consequently it is completely impossible to determine how useful this or that bit of *knowledge* will be. As a matter of fact, it cannot be doubted for a moment that what has contributed the most to human progress, even of a purely practical nature, has been disinterested research leading to theoretical knowledge, such as the invention of the infinitesimal calculus by Newton and Leibniz.

Certainly if mathematics today took a step forward comparable to theirs, we would not be long in seeing in its wake a great number of inventions as yet unimaginable.

When we see an airplane fly through the sky, we are well aware that this powered flight would have been impossible without the lightweight motor. But the merit of all those who contributed to the creation of this engine is infinitesimal compared to that of Sadi Carnot, since their work was only an application of his. Yet Carnot never invented anything at all; he merely established a purely theoretical proposition concerning the efficiency of heat engines in general. In the same way, all the applications of electricity, so numerous and so important from the point of view of our action on nature, are based on the theoretical work of Faraday, yet Faraday had tried in vain throughout most of his long scientific career to invent anything practical at all.[3] We can safely say that scientists worthy of the name have always been keenly aware of this situation. Comte himself, so well versed in the scientific knowledge of his time and endowed with such a keen scientific instinct, was at heart far from underestimating it; thus he pointed out, following Condorcet, that when the sailor determines his position, he profits from mathematical discoveries made by Greek geometers who lived twenty centuries ago and certainly could not have foreseen any utilization of this sort (CPP 1:53). If the founder of positivism nevertheless left these considerations out of account in constructing the model of his organization, he must therefore have been motivated by a very urgent need, and this can easily be recognized as nothing other than the need he saw to maintain the stability of science.

255. *The Stability Postulated by Comte*

It seemed to him that this had to be, above all, a stability of laws. For him, of course, it is laws that constitute the most essential part of science, since hypotheses themselves are never supposed to deal with the reality of things, but only with undiscovered laws. There can be no doubt, however, that if his reform could actually have been instituted, it would inevitably have brought about the immutability not only of laws but also of explanatory hypotheses. In fact, law is impossible without hypotheses concerning reality; it necessarily includes such hypotheses, because relationships can only operate between real substances. Kepler's laws are quite obviously founded on the Copernican hypothesis, and they are absolutely inconceivable without it. Just as striking an example is provided by the Newtonian law of attraction, which, in Comte's view, was the very model of definitive scientific truth. Certainly it is above all inseparable from the notion of space, that is, the assumption that space is really what our immediate perception takes it to be. If, on the contrary, we assume that the universe conforms to the general theory of relativity, the law is no longer anything but approximate and the anomaly of Mercury appears. Furthermore, let us recall that Comte, while claiming to rid science of assumptions about the nature of things, did not carry out this project very rigorously. He actually excluded only hypotheses consciously presented as such; unconscious hypotheses, and particularly those that form the concept of the world of common sense, remain within the bounds of science (§19). It should also be noted that the scientist actually clings a great deal less tenaciously to the immutability of his laws than to the immutability of his explanatory theories; the disappearance of theories is most often marked by a violent battle, while laws, without causing the slightest disturbance, vanish or take on the status of mere approximations (§55).

256. *His Error is the Same as Kant's*

Comte was imbued with the most profound admiration for the science of his time as it had been created by Newton and the Pleiad of great French scientists at the end of the eighteenth and beginning of the nineteenth century. One might even say that he exaggerated the value of this body of knowledge, since he believed that it was definitive in its essential features. But this resulted above all from the fact that he entirely disregarded the distinction we made earlier between the prestige enjoyed by scientific principles during a given epoch and their perenniality: the uncontested and incontestable authority attributed to certain principles in his time seemed to him a sufficient warrant that they were to constitute the foundation of physics forever.

To see how natural this attitude was, it suffices to note that Kant had the same attitude toward Newtonian science. One need only open the *Metaphysical Foundations of Natural Science* to observe that many claims that seem completely outdated today are treated as if they were part of the unshakable foundations of human knowledge. But in the case of Comte this overvaluation and confusion had particularly grave consequences because of his social and political preoccupations. Comte was profoundly impressed with the example presented by the solid social structure of medieval society, firmly grounded on the unity of the Catholic faith. Thus when he sought a foundation upon which to build his system, he became convinced that he could do no better than to emulate this model.

257. *The Intervention of the Secular Arm*

However, he was too much a son of the Revolution to consider that it was possible simply to return to this unity, and the unity of scientific beliefs, which was so manifest, seemed to him to provide a ready-made substitute. Nevertheless, even during his lifetime, cracks began to appear in the edifice as he conceived it; divergent tendencies appeared from different directions, and Comte could not have been unaware of them. Thus, to mention only one example – one, moreover, on which he himself laid great stress – the perfect exactitude of Mariotte's law, which seemed to him so primordial and so definitive in the simplicity of its mathematical expression, was brought into question by more precise determinations of vapor densities. To Comte, works of this sort appeared to be purely teratological excrescences which could not fail to be eliminated in the future development of science. In the meantime, however, they might well disturb the layman, if not the initiate. Thus, starting from his exaggerated idea of the stability of established science, he arrived at the idea that it was desirable to increase this stability even further by external means. The church militant and particularly the church triumphant had unscrupulously and ruthlessly made use of the secular arm against heretics in the interest of the unity of the ancient faith. Why would not the champions of the new faith, those entrusted with the authority of the "true speculative regime" of the future, act in the same way toward a Regnault who, moved by an "ever vain and gravely disruptive curiosity" and a "puerile curiosity stimulated by a vain ambition," dared attack Mariotte's law? As late as 1878, at a time when the great importance of this research – which formed one of the bases for the admirable structure of the kinetic theory of gases – had long been recognized, one of Comte's faithful disciples called Regnault, and very logically so, a "factious academic."[4]

258. *This Teaching Remains Without Echo*

Nevertheless, and this is important to note, the author of this last diatribe was not a physicist and, as a matter of fact, had only rather remote connections with science. Needless to say, the influence of Comte's teaching was in general extremely powerful. It still endures, and one need only open a book of science, or even of the philosophy of science, to see how careful the author generally is to show that he has not strayed from the principles of positivism. In view of this, one cannot fail to be struck by the fact that the side of the doctrine we have just discussed, which Comte himself considered so essential, nevertheless remained almost without echo among scientists. Despite his injunctions and his anathemas, physicists, chemists and biologists did not pause for a moment in their search for hypotheses quite obviously dealing not with lawlike relationships, but with the way in which phenomena are produced. When, only a few short years after Comte's death, work began on the chemical composition of the stars as a result of the discovery of spectral analysis, the whole scientific world welcomed the results with the greatest enthusiasm, and no one, absolutely no one, thought of pointing out that they were trespassing in an area Comte had most explicitly forbidden them to enter. Everyone also continued to use the microscope. As two very competent physicists have pointed out, today this research instrument is considered to have contributed probably more than any other instrument to the whole of science in all its branches.[5] Its use has been extended, notably by the splendid invention of the ultramicroscope, well beyond the limits that confined it in Comte's epoch, limits that already seemed to him to have been stretched too far. In still other areas, as they have perfected their research instruments, scientists have applied themselves to a closer and closer study of phenomena likely to upset previously formulated laws. Finally, no physicist worthy of the name has allowed himself to be upset by the misfortunes of Mariotte's law, which at the present time is generally considered merely an approximation. This whole side of Comte's doctrine has fallen so far into oblivion that scientists who believe themselves to be positivists often exhibit genuine surprise, not to say a little incredulity (we have had this experience more than once), when one attempts to call this to their attention.

259. *What can be Concluded From This*

This anomaly seems to allow only one explanation: The requirements laid down by Comte and the constraint he meant to impose on science were so incompatible with the true nature of science that scientists, in spite of the great weight carried by the name of the philosopher, did not for a single moment think of conforming to them.[6] And since, on the other hand, we know that these requirements and this constraint had no other aim than to transform science into a surrogate for religion, it would seem legitimate to assume – for even the errors of the great mind are quite often instructive – that it is here that the principal differences between the two domains can be identified. This can be quickly established.

260. *Theological Reasoning*

Let us first dispose of the quite widely held opinion that the path followed by theological reasoning is essentially different from the one followed by both scientific and philosophical reasoning. This is indeed in essence what Auguste Comte seems to have believed when he formulated his famous law of the three stages. It is quite true, as we have pointed out many times elsewhere, that theological thought admits of a particular form of causality, assuming that the divinity can intervene directly in phenomena. Thus the divinity is considered to act by means of his free will, somewhat as we feel ourselves act, but with a power immeasurably greater than man's. But it is no less certain that, in cases other than those allowing the influence of this *theological causality*, the theologian uses his reason exactly the same way as the philosopher or the scientist, and also in the same way as the man of common sense. The theologian too, by the very fact that he reasons, seeks to make things rational, to demonstrate that his assumptions conform to reason, that is, to draw them from the very essence of reason. Furthermore, if this were not so, would it have been possible for theological and philosophical thought to be as completely intertwined as they are seen to be during the long period of the Middle Ages? And for anyone who has taken the least pains to understand the mentality of this epoch, there cannot be the least doubt on this point: theologians are as fully convinced of the validity of their demonstrations as any contemporary scientist can be of the proofs of physics. Thus, to take only one well-known example, Saint Anselm's ontological proof of the existence of God certainly appears irrefutable to them. And since we know, moreover, that it still appeared irrefutable in its essential characteristics to Descartes and Spinoza, it would seem that we ought not be too surprised at this position.

261. *Its Progress, According to Saint Vincent*

It would also be inaccurate to claim that theological knowledge is immutable in nature, that all idea of progress is excluded from it. In the preface we cited Pascal's splendid image in which all of humanity is compared to "one single man who lives on and on and learns continuously." Attempts to find the sources of this image have revealed that it existed in Francis Bacon among others. But much earlier than these thinkers Saint Vincent of Lerins used it, and it is at least probable that later writers were inspired by him (Pascal may even have borrowed from him directly). Now, what Saint Vincent compares to the intellectual and bodily growth of the individual is "the intelligence, science and wisdom" of the whole church, which must increase over the ages.

262. *The Limit Imposed on this Progress*

At the same time — and it is here that we grasp how theology differs from science — Saint Vincent insists just as vigorously that faith must leave all its essential principles intact as it progresses. "Although the condition and appearance of one and the same man change, he nevertheless remains one and the same nature, one and the same person"; in the same way, there "must really be progress of the faith and not change." Thus the perpetuity of the dogma is established, and Saint Vincent is well aware that this requires as an inevitable corollary the existence of an authority to decide what is to be believed. He qualifies as rebellion the fact that "the particular turns against universality, innovation against antiquity, the dissent of a single wandering soul or a small number of them against the assent of all Catholics or the great majority of them."[7]

It can be seen that this is quite close to Auguste Comte's position, and we now see even more clearly why his organization so closely resembles that of the church, with its dogmas and its hierarchy.

263. *No Immutable Dogma in Science*

If one but considers science and its evolution with an open mind, one will realize that it can tolerate no immutable dogma of the sort conceived by Saint Vincent and Comte. It does, of course, have elements that persist throughout all its transformations. These are, first of all, the traits common to all our thinking in general, that is to say, the strictly logical schemas, as well as the tendency towards identification in time and space. Finally there is that tendency distinguishing scientific from philosophical thought: the deep conviction that a reality exists in the form of a substratum. As we have seen, all these characteristics taken together are responsible for the quantitative, and more particularly the spatial, nature of scientific theory. But everything else is most certainly subject to modification. What persists can be defined as the form of science and what changes as its content. However, it must then be remembered (as follows from what we saw in the preceding chapter) that this is not a distinction that we would have been able to establish *a priori* by reasoning, but one at which we arrive *a posteriori* by examining the evolution of scientific knowledge. For there is no doubt that, until very recently, three-dimensional space as it is known by our spontaneous perception was considered by scientists in general to be part of this immutable form of science, since it seemed to be one of the integral parts of our reason. Now relativism, or rather the fact that such a revolutionary hypothesis could have been accepted by a great many competent physicists, teaches us that this was an error. The only truly enduring element in our conception of space is the one we have characterized by our quotation from Borel (§28): an element qualitative in nature that distinguishes geometrical concepts from purely algebraic ones and that at the same time is the reason why the spatial, unlike the purely quantitative, can be used to construct a representation and an explanation of reality. Indeed, one could not imagine a more complete refutation of the idea that the essential content of science is unalterable than that afforded by the example of this recent transformation. What would Comte have said if he had been able to know relativism? For him anything reflecting on Mariotte's law already constituted a crime against science. What would he have thought of a doctrine that shakes the most basic foundations of the physics of Newtonian gravitation? After all, this physics seemed to him to have been established for all time — so much so that he declared research into the cause of gravitation, which was so troublesome for physicists (§60), to be entirely otiose and antiscientific, using this example, which he obviously considered a case in point, to demonstrate that scientific knowledge was not to go beyond the establishment of law. Relativism would have been a veritable nightmare for him. He would have seen there something apocalyptic, the twilight of his scientific gods and of science itself. But it is only the twilight of any idea aspiring to transform science into something closely akin to religion. There is no possibility of establishing for science a system that is *catholic* in the sense of Saint Vincent's famous definition, that is, a body of truths believed "everywhere, always and by everyone." If people have sometimes assumed the contrary, it is precisely because they have confounded the different terms of this expression, wrongly supposing that if a proposition was believed by everyone, it followed that it always would be. But such is not the case, and if one absolutely had to find a parallel to the faith of the scientist in the religious domain, one would find it not in the faith of the Catholic, so little open to doubt and so immutable in its essential contours, but rather in the uneasy faith of the heretic (as Henri Poincaré has wittily pointed out).[8]

Thus scientific skepticism, the conviction that any scientific affirmation dealing with the nature of reality is subject to revision, appears fully justified. But how are we to explain that a vigorous dogmatism appears to be superimposed on it?

264. *The True Significance of Fresnel's Demonstration*

Let us consider the circumstances under which one particular scientific revolution took place, say the replacement of the corpuscular theory by the wave theory in optics. What is the significance of Fresnel's principal demonstration, the one that above all succeeds in convincing his contemporaries? He shows that light added to light is able to produce darkness, which was inconceivable under the Newtonian hypothesis, concluding from this that we are dealing with undulation. Does he demonstrate this directly? This is so far from his mind that he does not even define the nature of that which undulates, namely the ether, merely stipulating that the waves must be transversal. We know that he was not able to fill this lacuna, and that generations of physicists since then have struggled vainly to give us a more concrete idea of this universal medium, whose properties, as required by the wave theory, are completely contradictory.[9] And yet his proofs seem so decisive that physicists like Arago, who were formerly very attached to the corpuscular theory, yield immediately. This is because, in their eyes, the truly important point is the refutation of the old concept, while the establishment of a new concept free from contradictions is somehow of secondary importance.

This at first seems strange, but we need only recall what we have come to understand about the true role of theories in science to realize why this is so. Theories are designed to furnish us with a representation of reality replacing the one spontaneously presented to us by our direct perception. In this capacity they are indispensable for science, which could never forgo the assumption of a reality independent of the consciousness (although it seeks to make this reality as rational as possible). This is why a theory is really abandoned only when there is another theory to take its place. It also explains why physicists treat the old theory with almost unlimited indulgence until this successor presents itself (as we explained in §130). But as soon as the successor appears, the situation changes. Of course the mind always finds it difficult — one might even say painful — to replace a familiar concept by a new one, since this involves a modification of its representation of reality (§55). Consequently, such an event is generally characterized by a sort of violent convulsion in the evolution of science; in an appendix to our preceding book we attempted to describe the upheaval accompanying the establishment of the antiphlogiston theory (ES 2:386 ff.). But it is nonetheless certain that it is henceforth the new hypothesis that will enjoy all the indulgence withdrawn from the old one, the partisans of the new doctrine being the first to make the change, followed later, after it has won out, by scientists in general. It goes without

saying that the new theory must not be too much weaker than the old one
in those areas where the latter had seemed sufficient, and, most importantly,
it must be able to explain that for which the other was unable to account.
But, beyond that, when it comes to the question of whether the new theory
itself is a really logical and coherent hypothesis, we shall not be too demand-
ing. For we need a reality, and once the old one has been destroyed, our
minds are predisposed to accept a new one, as long as we are able to consider
it, if not really good, at least better than the former one.

Thus what Fresnel said in effect was something like the following: I
demonstrate to you that light cannot be an emission of particles. On the
contrary, it behaves in many respects as if it were a wave. Do not ask me to
prove that it is one, nor even to describe precisely how the medium I postulate can vibrate in such a way. But since we absolutely must have a picture
of what is happening in space when we receive a sensation of light, and since
the old picture is no longer adequate, let us accept this new representation
until further notice.

265. *The Significance of the Copernican Demonstration*

This is also the real significance of Copernicus's demonstration: he proves
that it is impossible to suppose that the earth is motionless with the sun and
all the stars revolving around it. And since the astronomer must somehow
imagine how the celestial bodies really move, Copernicus offers as an alternative his hypothesis of a motionless sun with planets revolving in circles
around it. However, he does not dream of claiming that the motions are
exactly as he describes them. And his contemporaries are so much of the same
mind that when Kepler later shows that the trajectories of the planets are not
circles but ellipses, no one sees this as anything other than an extension of
the Copernican theory. Everyone considers Kepler a disciple of Copernicus,
including Kepler himself. And as for us, when we recognize − contrary to the
assumption of Copernicus and Kepler − that the sun moves continually, we
in no way think that we are destroying their system.

266. *The Significance of the Relativistic Demonstration*

The same is true for relativity. What is really demonstrated is the inconsistency
of the old assumptions: time and space cannot be what we believed them to
be since if they were, the Michelson−Morley experiments would have given
different results, light rays passing in the neighborhood of the sun would not
have been deflected in the way Einstein predicted, etc.

On the other hand, these observations can be explained, according to Einstein, if one grants that the universe has the spatio-temporal structure he described. But here again the innovator does not claim to be directly demonstrating that this structure is exactly the one he proposes. Indeed, we have noted that Weyl and Eddington attribute to the universe a more complex structure than the one suggested by Einstein. Now, in so doing they do not in the least take themselves to be destroying the Einsteinian theory, nor even, in general terms, to be opposing it. On the contrary, they consider themselves disciples and continuators, and Einstein himself obviously does not disown them as such, even though he does not entirely accept their results. Thus, in putting forth his hypothesis, he did not intend to proclaim a definitive truth, but rather to leave the door open to new theories. He did not claim that the universe had exactly the structure he attributed to it; he merely attributed to it the least complex structure that seemed compatible with the observations.

267. *The Name of the Theory Once Again*

In this respect, it can be said that the name of the theory is not too badly chosen (contrary to what we suggested in §53), for it refers to what is in fact the most essential aspect of Einstein's hypothesis: the necessary modification of our concepts of space and time. It is thus the same sort of designation we find in the case of *antiphlogiston* theory; this name, which originated with an opponent, was soon universally accepted, probably because both sides had the feeling that Lavoisier's innovation consisted primarily in the claim that phlogiston, until then the chemists' favorite way of explaining obtrusive phenomena, actually did not exist.

268. *New Facts and New Theories*

It goes without saying that the role of observations, here as everywhere in science, is to serve as points of departure for reasoning. It is because we want to use them to create a concept of totality (§15), a coherent representation of reality, that we so ardently pursue the consistency that constantly eludes us. This explains why the substitution of one theory for another is frequently the result of a new fact or series of facts.

This was true of Fresnel's theory, for example. But such is not always the case; on the contrary, it is perfectly possible for the new theory to afford only a better, simpler explanation of previously known facts with which the preceding theory had dealt after a fashion. Thus Copernicus, as we know, established his system using only observations well-known to his contemporaries, which they had fitted into the framework of Ptolemy's theory. Copernicus merely showed that they fit even better, with incomparably fewer complications, into the framework of his system.[10]

The case is analogous for general relativity, which, as we have noted (§216), also developed not as the result of new facts (which occurred only after the event), but from considerations based on the explanation of known facts (particularly gravitation). The same, moreover, had been true for Galileo's and Descartes's principle of inertia, based on the explanation of projectile motion, which had been familiar to man since his origins. Thus the two theories resemble each other in yet another respect.

269. *The* Convenient *and the* As If

Henri Poincaré, as we have seen in §§16 and 250, declared that if we choose a new hypothesis over an old one, it is because we find it more *convenient*. This could not be put better, for the new hypothesis brings together aspects of reality for which the old one could not account. It was certainly more convenient, from the time of Copernicus on, to describe the movements of the celestial bodies by the heliocentric hypothesis, and Tycho Brahe, even though he rejected this point of view (on mechanical grounds and also, as we pointed out in §252, because of the theological implications it seemed to entail), nevertheless was honest enough to acknowledge this advantage of the new concepts. However, if one compares his attitude to Kepler's, one realizes that one of the reasons Kepler, using Tycho's observations, arrived at results his teacher was unable to achieve, was the important fact that Kepler believed in the truth of a theory that Tycho saw only as a calculating device.

Thus, even more than the scientist who explicates established results, the research scientist must prefer the new theory to the old one. If he does not do so, if he is inclined to adopt the older position concerning the nature of reality, he cuts himself off from the possibility of a more complete understanding of this reality. The seventeenth century astronomer who rejected the heliocentric theory was forced to return forthwith to the assumption that the earth was immobile, in other words, to adopt the Ptolemaic explanation for the movement of the stars (unless he preferred the still more difficult explanation of the Eudoxian spheres), and clearly neither Kepler's laws nor, *a fortiori*, the Newtonian simplification could have been achieved working from these theories. Of course one can say that, for the research scientist as well, the new theory is more convenient, but in this case Poincaré's term, while remaining accurate, is clearly insufficient. For this reason it has been apt to give rise to misunderstandings, as is shown by the debate that arose after this word was applied to the geocentric and heliocentric theories. In our opinion, it would be better to use the term more *true*, which brings out more clearly that what is behind any explanatory theory of science is the need to presuppose a reality underlying the phenomena and the conviction that such a reality exists.

Joseph Bertrand, in speaking of Osiander's preface to Copernicus's *De revolutionibus*, where the heliocentric theory was presented as a mere calculating device unrelated to reality (probably to protect the work in some measure from the wrath of the Church), rightly said that "these lines, in which prudence poses as skepticism, are the negation of science."[11] There can be no doubt that these words express the sentiment of any true scientist on this subject.

Nevertheless – and this is very important to recognize – in asserting that the new representation of reality is more true than the old, science is not maintaining that it is unqualifiedly true; it is not claiming to have actually seized reality itself. What it offers instead is only a representation that, in some respect, must resemble this reality, a reality that in its totality remains essentially unknowable. In speaking of Fresnel's theory above, we made use of the locution *as if*, which actually characterizes this aspect of scientific theories rather well. A contemporary German philosopher, Vaihinger, used it as the basis for a whole system, which he calls the *philosophy of as if*.[12] But here again it must be noted that the expression, at least insofar as science is concerned, remains accurate only if it is not interpreted in a purely phenomenalistic sense (as Vaihinger himself did).

270. *Negative Dogmatism*

Thus, despite appearances, scientific theory basically is not and cannot be dogmatic in what it affirms; it is dogmatic only in what it denies. But it is easy to see how we could have mistakenly believed that it is the affirmations that are dogmatic and durable in a newly adopted theory. This is because once the old reality has been destroyed, the scientist is constrained to adopt the new position, whether in explicating this area of science or in pursuing his research. However, he only accepts it for lack of something better, in short, provisionally. This also explains why science displays so little tolerance toward those who try to go back to outdated theories. The scientist is keenly aware that these theories were abandoned because it was impossible to make

them agree with new facts, or because a new theory offered a better framework for known facts.

271. *The Scientist and the* a priori *Component of Science*

Finally, this helps explain why the scientist is apt to underestimate what
persists or eternally recurs in science, such as atomism or the reduction of
reality to space. He knows that both ancient atomism and the Cartesian
system had to yield, when the time came, to truer theories, and this alone
would suffice, if necessary, to make him consider blasphemous any attempts
to connect the work of Perrin to the theory of Leucippus and Democritus
or the work of Einstein to that of Descartes. He is of course pushed in the
same direction by the positivistic epistemology, the tendency to assume that
everything of value in science is experimental in origin (§214) and that from
this standpoint his methods are entirely different (for the better, to be sure)
from the theological or metaphysical reasoning of his predecessors. He is
undoubtedly wrong here, but just as his inclination to follow Bacon and
Comte by overestimating the value of experience is based on a sound conviction of what constitutes the distinctive character of science (§197), so also
his resistance to demonstrations attempting to connect him to a past he finds
generally outmoded is perfectly explicable. Moreover, it is also explicable
from a psychological point of view.

Indeed, it is not really his job to search for and determine the nature of
the permanent element of which we have just spoken. This is the job of the
philosopher, or more exactly, the epistemologist, whose work is carried out
by procedures the scientist is not accustomed to using. It would certainly be
just as well if the scientist then became conversant with the results achieved
by this method. Man does philosophy as naturally as he breathes (§52), and
the physicist, in reflecting on the nature of his work, is only too often moved
to do it badly. As far as the work itself is concerned, however, his ignorance
of the principles that really govern it is no great disadvantage, for his scientific
instinct is, in general, sufficiently sound.

Guided by this instinct, the scientist reconciles his skepticism with his
dogmatism — and rightly so, for the essentially negative character of the latter
allows it to be completely at home with the former.

272. *Mach's Attitude*

We observed earlier that Auguste Comte had misunderstood this situation by confounding these two opposing tendencies. It is quite interesting to see that the same confusion has reappeared today in a man whose way of thinking was, moreover, similar in many respects to that of the founder of positivism. The reader will have guessed that we are speaking of Ernst Mach. Mach does not seem to have read Comte; thus one would search in vain for Comte's name in the list of thinkers with whom he claims kinship in one of his last publications. The point of departure for his general theory, as well as the way in which it is developed, allows us to assume that his thought developed independently. Nevertheless, the similarity between his epistemological systtem and Comte's is so striking that even in Germany Mach's position has come to be called 'positivism'. Like Comte, Mach tends to confuse scientific belief with theological faith. However, instead of concluding, as Comte does, that the scientist must strive to stabilize his beliefs in order to transform them into immutable dogma, he draws the conclusion that in matters of science one is never constrained to conform to any particular opinion, even if it is the unanimous opinion of competent scientists in the area in question.

We see here a sort of absolute skepticism, growing out of Mach's strong resistance to the rebirth of atomism that occurred in science toward the end of the nineteenth century. Moreover, this attitude was consistent with his strict positivism, which inspired in him a horror of all explanatory hypotheses. When Planck reproached him for contradicting concepts that all physicists of that time considered firmly established, he replied that if physicists claimed that the reality of atoms was an essential point in the *credo* of their science, he would stop thinking as a physicist and would do without the approval of scientists altogether. Furthermore, it should be noted that in making these declarations, Mach explicitly invokes the sort of freedom of thought that exists in religious matters.[13]

Thus Mach — starting with principles analogous to Comte's, and like him confounding the nature of religious truth with that of scientific truth, but turning his attention more toward hypotheses (while Comte concentrated above all on laws), and, moreover, convinced of the need for unlimited freedom in matters of faith (whereas Comte was convinced of the necessity of absolute stability and authority) — arrives at conclusions diametrically opposed to those of his predecessor.

But in no sense is freedom of thought the issue here. What Planck had intended to assert (and in this respect Comte is certainly closer to the truth than Mach) is that, just as one could not really make advances in astronomy in the seventeenth century without adopting the Copernican theory, nor in the eighteenth century if one denied Newton's concepts, and just as one could not profitably further the cause of chemistry in the first half of the nineteenth century if one ignored Lavoisier's ideas, so today one cannot do research on the constitution of matter without positing the existence of atoms. This is true because, at a given time and under given conditions, scientific faith is strong and generally accepted; it is completely futile to insist on freeing oneself from it, as Mach tried to do, by means of a general profession of quasi-philosophical skepticism.

Furthermore, in the case of the atomic concepts against which he meant to protest, one can see that here again the essential part of the theory is a negation: contrary to the assumption we once made, perhaps not quite consciously, an assumption that thermodynamics in particular had made explicit, it is impossible to consider matter to be infinitely divisible or uniformly constituted; on the contrary, we must postulate that it has a particulate structure, and the dimensions of these particles can be indicated. But one is careful not to maintain that these are the ultimate particles of matter. As a matter of fact, the physicist readily accepts the idea that what he arrives at through observations and calculations like those of Perrin concerning Brownian motion or Lord Rayleigh on the blue of the sky is, above all, only molecules, which are made of atoms, which in their turn, and in spite of the original meaning of the word, are considered to have a very complex structure.

273. Anti-Einstein and Pro-Phlogiston Theorists

Thus there can be no doubt that if all the physical criteria that have been indicated or will someday be imagined tend to confirm the theory of relativity, the scientist working in relevant fields — that is to say, dealing with motions of a sufficient velocity that the effect of the divergence between the new formulas and classical mechanics makes itself felt — will be forced to adopt Einstein's point of view. If he does not do so, he will put himself in the same position as the German chemists at the end of the eighteenth and the beginning of the nineteenth century who, because they refused to accept the antiphlogiston theory (largely motivated by a nationalistic prejudice, according to a most competent observer who is himself German) found themselves left behind and in some sense rejected by the mainstream of the chemistry of their time.[14]

274. The Attitude of the Layman

As for the interested layman, on the basis of the preceding discussion we would seem to be able to identify with some precision the point of view that will be required of him by relativity (always assuming, of course, that its triumph is sufficiently complete and durable). In his everyday life he will not be asked to give up his accustomed concepts of space and time. If we but ask ourselves, we realize that we do not ordinarily abandon our concept of the immobility of the earth, and that the very consequences of its sphericity may even seem paradoxical to us (§231), which obviously proves that we hardly ever think about this long-accepted truth. Furthermore, even when observing particular phenomena more closely, one can completely disregard all relativistic concepts in the great majority of cases and continue to use familiar concepts of space and time. In speaking of relativism, Borel aptly says:

Architects, like all civilized men, must have heard since childhood that the earth is round, but when they build a house, they have no need of this truth, and, on the contrary, must be persuaded that the vertical lines are parallel.[15]

It is thus only with regard to very particular and very subtle phenomena that one must keep in mind that space and time, contrary to one's firm convictions, do not constitute the unchanging framework for reality, nor even a rigid form necessarily imposed on reality by intuition (in accordance with the Kantian conception). They are instead, at least in some respects, part of this reality itself, so that the way we conceive them can to some extent be modified as science progresses, in the same way we have modified many other apparently unassailable beliefs about the nature of things.

275. Useless Excesses

But if someone who admits that he himself is a layman in this field can nevertheless be permitted to give advice to experts, perhaps it would be in order for the relativists on their side to consent not to frighten away the ordinary man any more than absolutely necessary. What we mean to say is that they could leave philosophy to the philosophers and remember that it serves no useful purpose in matters of science to seek to "enlighten by provoking us." These are the words in which a well-known Hegelian very aptly characterized the method of his master.[16] The method was not without drawbacks in the hands of Hegel himself, but it also offered advantages. Indeed, philosophical theories are by nature rather fluid and nebulous, and the author is only too tempted to blur their outlines to make them more easily acceptable to the reader. Consequently such presentations often unabashedly display an easy and somewhat shallow syncretism, and in order to combat this propensity, it may occur to a philosopher to express a concept in its most extreme form, the form most opposed to everyday common sense. In the realm of mathematical formulas, however, everything is rigor and precision. Once the meaning of each symbol has been determined by qualitative interpretation (§116), the proposition says exactly what it must say. There is therefore no need, in translating the mathematical formula into a verbal formula, to give it a purposely paradoxical form. Thus, to cite only one example, it is not clear what is to be gained by asserting that the four dimensions of the universe are perfectly isotropic, meaning that the temporal dimension is not distinct in nature from the spatial dimensions. This is a very interesting proposition from the psychological and epistemological point of view, for, as we explained in §71, it reveals the most profound tendencies of relativism. Nevertheless, this proposition is extremely shocking not only to the man of common sense, but also, and perhaps even more so, to the physicist, to whom Carnot's principle offers a clear and unshakable conviction of the reality and importance of irreversible becoming. It would surely be better in this case to evoke Langevin's declaration, as accurate as it is emphatic, that it makes "no sense" to say that "time is a fourth dimension of space" (§66). As a result, relativism would tend to appear less like a deformed monster — a teratological being, the fruit of some sort of scientific aberration — and more like one of the normal offspring of science, which in reality it is.

NOTES

[1] Tychonis Brahe, *Epistolarum astronomicarum libri* (Uranienburg and Frankfurt, 1610), p. 190. Anatole France, in a well-known passage, has pointed out very aptly and colorfully the principal differences between medieval man's idea of the universe and our own (*Le jardin d'Epicure*, Paris, 1895, p. 1).

[2] CPP 1:298; 2:11; cf. pp. 6–8; 6:637–638; cf. 3:369, 6:596.

[3] Cf. Wilhelm Ostwald, *Grosse Männer*, 2nd ed. (Leipzig, 1910), p. 115.

[4] CPP 3:369; 6:586, 637–638; Pierre Laffitte, 'Cours de philosophie première – Première leçon,' *Revue occidentale* 1 (1878) 288.

[5] Aimé Cotton and Henri Mouton, *Les ultramicroscopes* (Paris, 1906), p. 1.

[6] It will perhaps not be out of place to recall here that an actual attempt to *organize* science, a very modest attempt compared to Comte's designs, ended in deplorable failure. Colbert, with the best of intentions, had thought he could prescribe to the members of the *Académie des Sciences* how they were to go about their work. Gaston Darboux certainly expressed the unanimous feeling of scientists toward this misguided effort when he wrote: "This organization did not turn out well, as will easily be imagined. Nothing is more deadly than restrictions on the freedom of the scientist. Research must

be free and the spirit must be able to blow where it listeth" (*Eloges académiques et discours*, Paris, 1912, p. 299).

[7] St. Vincentii Lirinesis, *Commonitorium Primum*, Chs. 23, 27, in *Patrologie Migne, Patres Latini* (Paris, 1846), Vol. 50, pp. 667, 674.

[8] Henri Poincaré, *Savants et écrivains* (Paris: Flammarion, s.d.), p. 6.

[9] Sir Oliver Lodge, in 1908, while emphasizing the need to assume the existence of an ether, nevertheless concluded that Lord Salisbury was correct when he wittily defined it as being "little more than a nominative case to the verb to undulate" ('The Aether of Space,' *Nature* 79 [1909] 323).

[10] This explains the at first puzzling fact that the heliocentric system, though perfectly well-known in antiquity (cf., for example, Heath, *Aristarchus of Samos*, Oxford, 1913, *passim*), was so little able to gain acceptance that it subsequently fell into complete oblivion. It is because the only reasons that militated in its favor were reasons of simplicity and (until the introduction of the principle of inertia) they came up against mechanical considerations. These reasons of simplicity became all the more persuasive, however, as the number of observed facts increased, and it became necessary to increase the complications of the reigning theory to account for the anomalies that these facts seemed to present for it. As early as the Middle Ages these complications had reached the point that they were apparent to any unprejudiced thinker, as is illustrated by the well-known anecdote concerning Alfonso X of Castile. The king, who was much enamored of astronomy (we owe to him the famous Alfonsine Tables) is said to have cried out one day, when confronted with the intricate pattern of epicycles, that if he had been in the Creator's place, he would have arranged things more simply. Even if we assume that the quotation is apocryphal, like so many other historical statements, the fact that it could be invented nonetheless shows that this was a sentiment that seemed quite natural to his contemporaries. Thus it was that the Copernican theory was finally able to triumph despite the formidable aid given to its adversaries by theological considerations (cf. §252), which, Duhem to the contrary (*Le système du monde*, Paris, 1913, 1:425), certainly played only a very small role in antiquity.

[11] Joseph Bertrand, *Les fondateurs de l'astronomie moderne*, 3rd ed. (Paris, 1865), p. 51.

[12] Hans Vaihinger, *Die Philosophie des Als-Ob*, 2nd ed. (Berlin, 1913).

[13] Ernst Mach, *Die Leitgedanken meiner Naturwissenschaftlichen Erkenntnislehre* (Leipzig, 1919), p. 11.

[14] Hermann Kopp, *Geschichte der Chemie* (Brunswick, 1845), 1:341, 345.

[15] Emile Borel, preface to Einstein, *La théorie de la relativité restreinte et généralisée*, trans. J. Rouvière (Paris, 1921), p. 8.

[16] James Hutchinson Stirling, *The Secret of Hegel* (Edinburgh, 1898), p. xlix.

CHAPTER 25

THE OUTLOOK FOR THE FUTURE

276. *Philosophical Reasons for a Return to Ordinary Time and Space*

Does an outlook for the future emerge from our analysis? In the course of this work we have often pointed out that the fate of relativism appears to us to depend, above all, on future experimental observations. We fully appreciate that this precludes any effective action by philosophical criticism in this domain, and such a position may seem excessive. In particular, it will be asked whether certain circumstances we have brought to light here do not justify the belief that there are other reasons besides experimental ones for physicists to return without delay to the concepts of space and time that characterize our immediate perception.

Indeed, it will be said, the first thing reason postulates in seeking to explain a phenomenon – to show that it conforms to the canons of reason – is, as we know, the concept of the identical. In order to adapt the identical to the diversity of sensation, reason diversifies it by means of the notion of space. But this notion itself, as it is presented by common sense, seems to us to be simple. Now it is no longer simple in relativism. Can one claim then that this space, with its complications, its *curvatures*, really explains phenomena? Certainly many contemporary physicists say that it does, claiming that the relativistic explanations satisfy the requirements of *their* reason. But is it quite clear that they are not the victims of an illusion? The age-old practice of science has accustomed them to consider that spatial explanations are the most, indeed even the only, satisfactory ones, the only ones having at least the appearance of being definitive. It is therefore not too surprising that they are inclined to find the relativistic explanations satisfactory, because of their unmistakably spatial character. But, in the end, they will realize full well that here the term *space* signifies something very different from what it has designated in physics until now and specifically that the new notion includes an element of diversity that was entirely foreign to the old one. In the final analysis, they will come to see that a 'particular structure' of space differs only verbally from the presence of a differentiated matter in homogeneous space, and consequently they will more or less unconsciously but inevitably seek to return to the identity

238

of homogeneous space. This tendency will operate surreptitiously, so to speak, as it has done elsewhere in physics. Identity will tend to be realized, in theories that are the expression of it, through experimental observations which — one can at least hope — will finally show reason's postulates to be justified here as elsewhere. In our treatment of the relationship between the *a priori* and experience in Chapter 22 we explained in somewhat more detail how a change of this kind must be understood.

277. *Is the Gain Proportional to the Sacrifice?*

This argument can be put in a somewhat different form, one that is more abstract and will perhaps reveal more clearly how it is connected to some of our conclusions.

In order to take the form relativism imposes on it, scientific reason must obviously do violence to itself to some extent (§243). It consents to a painful sacrifice, doing so only for the sake of the advantage it hopes to gain by extending its domain. The sacrifice and the gain must be proportional, however, or rather, given that it is a question of profoundly modifying our representation of what exists — a process man naturally resists — the gain must be worth immensely more than the loss. Is this the case here? Can it not be claimed, on the contrary, that the gain in the extent of the rational domain brought about by relativism is inconsequential compared to the effort it requires of our imagination? To use a current expression, is the game here worth the candle?

Let us note that this is the same sort of argument referred to by those adversaries of relativism who criticize the narrowness of its experimental bases (§76). We have shown that this criticism is not justified, that Einstein's theory is actually based on Maxwell's equations and the Lorentz transformations, that is to say, on the very foundations of the present physics of electricity, which tends to include the whole of physical phenomena. But even taking this into account, it is possible to believe that relativism is just a transitory phase and that science will eventually return to procedures more like those it has always followed.

278. *Can an Element that has been Abandoned by Reason be Reinstated?*

We saw in the preceding chapter that the evolution relativism requires of reason is essentially to make it repudiate a postulate that seemed an integral part of it. And if, after this relativistic change has taken place, the change were to be reversed, this would obviously be equivalent to reinstating the element in reason.

Is such an occurrence possible? We frankly do not know and find it extremely hard to make a judgment. The actual examples of an evolution of reason are few in number, as we have seen, and it would certainly be presumptuous to attempt on this basis to deduce general laws permitting us to make predictions. All one would be justified in concluding here is that while reason appears to change only rarely and with difficulty, it seems always to have gone in the same direction. We know of no example where a change took place in the opposite direction, that is to say, where a postulate once abandoned by reason was subsequently reestablished. One obviously cannot use Hegelian becoming as an example. Hegel did of course claim to have made becoming conform to reason, and we have observed that the modification he required of reason went altogether in the same direction as that required by relativism (§246). On the other hand, it is also obvious that at the present time this concept is not taken to be valid, that the intellect governing the reasoning of our contemporaries seeks the identical, following the tautological pattern of our forefathers that so horrified the author of the *Phenomenology*. Can it be claimed that there was an interim period when the Hegelian theory actually prevailed? We believe we have completely established the contrary: even those disciples of Hegel who were most faithful to the spirit of his doctrine did not succeed in reconciling themselves to the excessive demands it made in this respect (ES 2:63 ff.). However little inclined one is to exaggerate the success of relativism, one must nevertheless admit that it has a much firmer hold on the minds of its adherents than did its Hegelian counterpart. Thus, any change that would bring about the abandonment of relativity theory and a return to the traditional concepts of space and time would be unprecedented.

279. *As Science Progresses it Includes More of the Irrational in its Explanations*

Returning to the argument in its initial form, we can likewise observe that, as paradoxical as it might seem from the philosophical point if view to transfer diversity to the concept of space itself, this nonetheless conforms to the general progression of scientific thought. The opponents of atomism have quite often accused this doctrine of indulging in a sort of sleight of hand by simply endowing the particles with the properties the theory was supposed to explain. Obviously this reproach was largely unjust, and they were thereby underestimating what Poincaré called "the power of grouping,"[1] in other words, the explanatory power inherent in spatial representations. But there was nevertheless something true in their accusation. Indeed, although reality becomes more and more accessible to our explanations, more penetrable by reason, as our knowledge of reality progresses, at the same time we also acquire a deeper and deeper conviction that it is not entirely explicable, that there is something there that resists this penetration. This is why science, as it advances, becomes more tolerant of this irrational element, more inclined to include it in its explanations, that is to say, in what it seems implicitly to consider part of the rational.

280. *The Methods of Philosophy and of Science*

The philosopher would certainly be inclined to consider that science is sometimes overly tolerant in this respect, too disposed to console itself, as Planck does, with the reflection that

were we in a position to answer all these questions satisfactorily [questions such as the speed of light, the gravitational constant, the mass of the electron, the quantum of action], physics would cease to be an inductive science, and it will certainly always be that.[2]

The philosopher will say that when the physicist speaks in this way, he is forgetting the principles underlying his research; he is losing sight of the fact that the purpose of research is to establish a coherent representation of reality, which can obviously be accomplished only if nature shows itself to be intelligible in the sense given this term by d'Alembert, Wien, and Planck himself in the passages cited in §11. But it is at just this point that the philosopher and the scientist face the fundamental problem that divides them. Science, having grown out of philosophy, is still today only a particular type of philosophy (§195). Just like the branch of knowledge we officially call philosophy, science seeks to know reality. But its method is different from that followed by philosophy strictly speaking, in that it vigorously affirms the existence of a reality located outside the self as the cause of sensation. This is only a temporary concession on the part of science — or rather on the part of reason, from which both science and philosophy emanate — for, in the final analysis, science too attempts to explain *all* reality, which can be done only by deducing it all from the categories of reason. Nevertheless, this goal, immediate for the philosopher, is remote for the scientist. Since he feels vaguely but intensely that the contradictions, not to say incoherencies, at which he arrives are imposed by the very spirit of the method he follows, the scientist is generally impervious to the reproaches of the philosopher who makes him see clearly the defects of his theories. The philosopher is undoubtedly correct in claiming that from the standpoint of the complete and definitive deduction – of the *system* – these are redhibitory vices. The scientist, however, is deeply convinced that these anomalies spring from the very reality he is investigating – from the fact that this reality does not entirely submit to the imperatives of reason – and that in allowing them he remains closer to this reality, accommodating his reason instead to the imperatives of reality. This is why, in many areas, diametrically opposed currents can coexist in science: for example, science can at the same time, by mechanism and the principles of conservation, tend toward the negation of diversity and, by Carnot's principle and the concept of the irrational in general, tend toward the affirmation of this diversity. In a still more clear-cut contradiction, these two currents can be seen in the two successive operations by which we seek first to distinguish atoms and then to dissolve them into undifferentiated space (ES 1:216). It is not that the mind of the scientist does not, like that of every thinking man, aim at the greatest possible unity in the concepts he adopts. It is that the scientist, in his desire to get closer to nature than the philosopher does, to live in a more intimate communion with it, feels that this aspiration does not allow him to yield too much to the temptation of attributing to phenomena more coherence than they perhaps have, or at least of attributing to them a coherence that might not be theirs, because it obviously would be primarily the coherence of his own mind. Thus the scientist most often seems to be ignorant of the distant goal mentioned above, or to forget it so completely that he seems to be giving it up definitively – and that he can never do.

281. *The Explanations of Lucretius and Descartes and of the Moderns*

The separation between science and philosophy has obviously been accentuated as each has progressed. Thus, from the present point of view, modern science can be seen to compare unfavorably with the sciences of earlier epochs: one need only compare the beautifully simple explanations of Lucretius or Descartes to the infinitely complex modern theories with their molecules, atoms and subatoms. At each phase these theories introduce properties that scientists profess to accept only provisionally, but that are actually perfect examples of 'occult qualities'. Physical science, originally only a narrow stream, is today a powerful and raging river whose course is apparently irresistible. But the flowing water is no longer the same: then it was crystal clear; now it seems opaque with the silt it carries, silt which is, moreover, the inevitable price of its growth.

We must not be surprised then if the scientist has accepted the complications relativism entails for the concept of space; he is accustomed to doing this sort of thing and does not expect too much of the theories put to him.

282. The Complexity of Spatial Intuition

It should also be added that the argument based on the supposed simplicity of the Euclidean concept of space is perhaps in reality less strong than it first appears. Indeed, if the evolution leading to the concept of hyperspace teaches us anything, it is certainly that our spatial intuition is not simple. Basically, this already follows from the elementary observation that in order to establish a science of space we need a whole series of axioms or postulates, the whole set of which is required to construct the network of deductions that constitutes geometry. It was Lobachevsky and Bolyai, however, who made it clear that the axioms and postulates with which we are dealing are notions entirely independent of one another, since by leaving one of them out we can arrive at a system equivalent to Euclidean geometry in the sense that, like the Euclidean system, it is exempt from self-contradiction. And since, once again, this system is derived from Euclid's by eliminating a postulate, it is incontestable that from this point of view the underlying notion seems simpler. [More accurately, by *replacing* one postulate by another.]

283. The Scientific Reasons for a Return

Despite all this, there are still considerations weighing heavily in favor of the return of theoretical physics to the concept of space and time implied by immediate perception. Certainly this conception, which even the most dedicated relativist will necessarily use in the myriad circumstances of everyday life, will continue to exert a considerable influence on the way he thinks, and with the slightest excuse he will eagerly return to it.

Furthermore, apart from reasons connnected with spatial concepts, which are consequently geometrical in nature, one can identify others, more strictly physical, having the same effect. For as Wien very aptly points out, relativity theory also requires sacrifices in the physical domain, seeming to close off definitively certain avenues of investigation that were previously open to the scientist's curiosity. For example, one henceforth gives up any possibility of deducing the velocity of light, which becomes an ultimate given due to the very nature of space and motion.[3] Of course, we ourself have just observed how inclined scientific reason is to accept irrational elements of this sort and incorporate them, so to speak, into its explanatory apparatus, if the line of argument seems to require it. But we have also seen that reason does not forgo explanation once and for all; on the contrary, if the occasion presents itself, reason will necessarily tend to turn back against what seems to be part of its basic foundations and require that this too be explained and made rational. This is sufficiently demonstrated by the way in which, through the introduction of statistics and through the theory of relativity, physics succeeds in making rational (even if only partially) the notion of becoming, which it previously seemed to accept without protest as irreducible (§190). And it is thus quite possible that in the future an attempt will be made to return to an explanation of the velocity of light, assuming a transmitting medium, an ether, whose existence also seems to be suggested by the very mysterious and complex phenomena of black body radiation. Now as soon as ether has come back into science, even by the back door, the question will again inevitably arise as to whether motion with respect to the ether does not take on physical significance, as Wien does well to point out (WW 272).

284. *The Return Nevertheless Seems Unlikely*

But all these developments, of whatever order, will certainly depend primarily on the circumstances, that is, on the direction given to research and theoretical deductions by experimental observations. For, as the entire course of science demonstrates (and as as we have seen in more detail in Chapter 16 in connection with Bohr's theory), the physicist is, above all, preoccupied with extending the range of his explanations and encompassing the greatest possible number of observations; no sacrifice is too great if it allows him to satisfy what might almost be called his passion. Wien, who (as we saw in §105) has criticized the overly complex nature of Weyl's theory, has also expressed similar concerns about the theory of relativity in general, although he has been careful to add that "if the consequences drawn from the theory are actually confirmed, this conception will remain of great importance nevertheless." And, speaking in more general terms of the objections raised by some philosophers against the use of statistical methods, this scientist declares that physics "will make use of every available resource, no matter how strange it may seem, and even if the aims of the theory of knowledge have to be redirected because of it" (WW 234, 280). Indeed, there can be no doubt that the physicist will not be turned aside in his pursuit of what he considers

the essential goal of science by scruples the philosopher might suggest to him.
Borel has said that

> it will only be when someone, starting from principles different from those of Einstein, succeeds both in foretelling new phenomena and in co-ordinating the old ones, that the new principles he uses will either take a place beside those of Einstein, or perhaps even replace the latter entirely (ET iv; Eng. xi).

Everything seems to indicate that he is completely right and that this is what the future holds. Consequently, one cannot help thinking that, despite the philosopher's scruples and the physicist's implicit preference for a notion of space devoid of curvature, it nevertheless seems highly unlikely that once the battle of relativism has been decisively won, the progress of physics will later be marked by a return to the prerelativistic concept.

285. Will other Geometrical Axioms be Abandoned?

Instead of such a return, perhaps we should even anticipate a progression in the opposite direction, that is, a continuation of the evolution that led from prerelativistic to Einsteinian physics. Einstein's physics, as we have noted, rests on a concept of spatial reason entailing the elimination of a previously posed postulate. Since all the axioms of a geometry are by definition independent of one another, one can predict with certainty that it will be possible to construct systems as free of contradiction as those of Lobachevsky and Bolyai by eliminating one axiom or another. Attempts of this sort have already been made (Borel, ET 52; Eng. 44).

As a matter of fact, the theories of Weyl and Eddington basically involve an enterprise of this sort. Quite recently Einstein too has made a similar attempt, while Cartan has treated the general problem of four-dimensional geometries and their application to physics.[4]

286. Will Spatial Continuity be Abandoned?

One foresees the same sort of possibility in a process that would lead to the elimination of the postulate of spatial continuity. In fact the physical data already exist: obviously we have in mind those on which quantum theory is based. We have spoken of the great importance these facts have assumed in contemporary physics, pointing out how contemporary theorists see them as shaking the very foundations of science. Let us give an example from a recent work by the author of quantum theory himself. The wave theory of light postulates that light emitted from a source spreads out indefinitely over greater and greater surfaces; this is one of the fundamental concepts of the theory and one that has not changed since Huygens. Now how can this concept be reconciled with the concept of an indivisible quantum?

To be fair, we must note that the contradiction here does not follow from the theoretical concepts alone, but is the result of conclusions more or less directly required by experimental observations. For example, when a metallic surface in a vaccum is bombarded with ultraviolet rays, electrons are ejected from it. Phillipp Lenard discovered that — contrary to what one would expect — the velocity with which these electrons are ejected does not depend in any way on the intensity of the radiation; it depends exclusively on the wavelength, that is, on the color of the light, increasing as the waves become shorter. Thus, no matter how weak we make the light that strikes the metallic plate, the electrons it ejects will always have the same velocity; only the frequency of their ejection will diminish. Therefore — as Planck himself points out — one must ask where the kinetic energy of the electron can be coming from. If one adheres to the concept of continuous waves, following Huygens and Fresnel, one would be led to assume at the very least that the intensity of light can be accumulated at the points where the ejection of the electrons is supposed to take place; however, this is a notion that seems to contradict the very foundations of the concept.

One can, of course, consider a return to Newton's theory of emission. But in that case, as Planck realizes, there looms the formidable obstacle of the observation that, in Fresnel's hands, decisively established the wave theory. For how — without infringing on the principle of the conservation of energy — are we to explain that two identical quanta of light meeting after traversing space independently are able to interfere, that is to say, to neutralize each other?[5] One can arrive at this explanation only by means of a whole series of assumptions that are extremely complex and, we must admit, shocking to basic common sense.

It is this fact that leads the relativistic physicist to ask himself whether the arbitrary element eliminated by Weyl and Eddington is really the last one that must be eliminated from his geometry. Do not some still remain, he asks, that keep us from completing our deduction of reality? In particular, why is it that so far all attempts to reduce the quantum of action to an effect of [the relativistic spatio-temporal] universe have failed, as have the attempts to represent positive or negative point charges in this manner? Is there not still a postulate underlying our calculations that we are justified in rejecting, namely the postulate of continuity? It is very nearly in these terms that

Langevin put the problem at the December 1921 *Congrès de philosophie.*[6] Borel is of the same opinion, because he thinks it would be necessary to "introduce this discontinuity [of the quanta] into geometry itself" and "to construct [on the subatomic level] an entirely different geometry [from the one we know] of a discontinuous character." Assuming that one succeeds in elucidating somewhat, "in the domain of mathematics, the problem of the abstract relations existing between the continuous and the discontinuous," one will perhaps be able to "hope to find in some of these relations a suitable model for representing and predicting certain phenomena," and in particular those summed up by quantum theory.[7] It can be seen that it is indeed a question here of accounting for physical diversity by transferring it to space, just as relativity theory does; one expects to accomplish this by abandoning a postulate of Euclidean geometry, in this case the postulate of continuity.

Those who are at all familiar with the scientific concepts of the past will note, moreover, that these ideas are not completely without precedent in the history of human thought: for the Pythagoreans space was a discontinuity formed of points, and later, in the Middle Ages, an Arabic school of philosophy, the Mutekallimun, attributed a similar nature to space (cf. IR 89; Eng. 86).

287. *The Constant Progress of Mathematical Rationalization*

All this obviously supposes that physical science will continue to develop in the same direction it has until now and, above all, that it will become more geometrical. Indeed, there certainly seems to be more or less continual and unswerving progress in this direction, by which– to put it in the most general terms possible – mathematical rationalization extends its domain over reality in general and becoming in particular. This has been going on for centuries, ever since human thought, through the Renaissance, reforged the chain of progress almost completely broken (Auguste Comte and other defenders of this period to the contrary) by 'the frightful adventure' of the Middle Ages, returning to the panmathematical tradition of science under the influence of Galileo and Descartes. And not only have all attempts to the contrary failed – whether they involved a return to the Peripatetic science of the Middle Ages or an effort to reinstate some form of quality in science – but panmathematicism has uncontestably become stronger with time. The contemporary physicist is more influenced by mathematics and more convinced that it is the only appropriate way to understand reality than was his predecessor a century ago. Moreover, the theory of relativity shows with great clarity how strong this tendency is, how far the scientist is disposed to extend his willingness to sacrifice in order to satisfy this overbearing inclination, and how little weight the most essential norms of common sense carry in the face of its demands.

288. *Can this Past be Projected into the Future?*

Given the natural human tendency to project the past and the present into the future, one would at first be tempted to conclude that this process will continue indefinitely, and that if, perchance, the theory of relativity should eventually collapse as a result of adverse experimental findings (an eventuality we considered in the preface), the scientific mind would find other expedients in order to continue the job of making reality more mathematical.

But the reader knows from our discussion in §121 how difficult we find all prediction of this kind. Indeed, how can one predict the paths reason will follow under conditions of which we cannot conceive even the barest outlines?

289. *The True Nature of Reality*

This seems all the more futile since we no longer have a very secure idea of the true nature of reality to guide us in this matter. When Plato, Galileo or Descartes proclaimed that reality could only be mathematical, they undoubtedly controverted the obvious fact (so perfectly brought out by Aristotle) that the mathematical is powerless to recreate quality. But they were nevertheless relying on the important fact of the absolute agreement mathematics reveals between intuition and reality. There can no longer be any question of intuition in relativistic mathematics, however, or at least not in the same sense as before. On the contrary, in the mathematics of relativity one must say that immediate intuition is somewhat defective and must be corrected in order to conform to empirical observations. How dare we claim then that the physical must coincide with the mathematical, given, first of all, that we sense – probably more clearly than our forefathers because of the very development of science – how much of reality mathematics fails to capture; given also that we have acquired more data concerning the existence of a given independent of reason, of which data those summarized by quantum theory are the most precise and the most threatening to a coherent representation of reality; and given, finally, that we have come to recognize the inadequacy of our mathematical intuition? The only support we have left is the fact of past success: physics has mathematicized, it has succeeded marvelously, therefore it will always succeed if it follows the same route. Anyone can recognize the weakness of such a deduction. Moreover, we are well aware that reality is essentially opaque and that all propositions concerning it retain something of the unpredictable. If this were not true, we would not need experiments, and recent discoveries have certainly brought home to us what painful surprises they hold for us.

290. *Extramathematical Deduction*

Therefore, the fact that extramathematical deduction has not succeeded up to this time does not really allow us to conclude anything about the future. Not so long ago Schelling and his school thought they could make advances in science using such a method and even prided themselves on alleged successes achieved by extramathematical reasoning. They were mistaken; the 'philosophy of nature' did not yield very much. It 'did not work', as we say in the laboratory in speaking of a theory whose application does not lead to favorable experimental results. But, after all, no valid demonstration allows us to conclude that this *will never work* and that the agreement between thought and reality will not manifest itself tomorrow in quite a different domain from that reserved for mathematics.

291. *Political Considerations*

It must be recognized, moreover, that the future orientation of science may depend in part on factors outside the domain of disinterested thought and research, on external — in a word, purely political — circumstances. In the nineteenth and at the beginning of the present century, it was possible to construct theories implying a human progress that is continual and in some sense necessary in itself, a progress independent of all external conditions, itself irresistibly creating the most favorable conditions for its development. But the Great War and its aftermath have destroyed this illusion. Given, on the one hand, the moral isolation and the growing hatred between peoples and, on the other hand, the rapidly increasing power of the means of destruction, against which any kind of adequate defense seems impossible, the disappearance of a large part of the immense body of knowledge acquired by civilization has become an imminent threat. Without going as far as the gloomy eschatology of Branly — whose imagination seems to be akin to that of the author of the *Dies Irae* — and without believing that man will be reduced to banishing any use of the intelligence and reverting to the state of the ant, social but apparently incapable of any progress,[8] it would seem that we have legitimate reasons to fear the coming of a new Middle Ages, especially since archeology has taught us that the intellectual reversal of the Middle Ages was by no means the only one humanity has had to record in the course of its destiny. It has now become quite evident that the splendid Mycenaean or Aegean civilization must have perished following a political cataclysm similar to the one by which the barbarians put an end to the Greco-Roman civilization. The former perished as completely as the latter, for the faint memories the Hellenes of the classical epoch retained of the Mycenaean civilization are, if anything, less precise than those the Middle Ages retained of antiquity. And we are now acquainted with vestiges — admirable for the artistic sensibility that inspired them — from a much earlier epoch, so ancient we can hardly reckon the number of centuries that separate us from it. This so-called quaternary civilization seems to have disappeared without leaving any lasting influence on the development of humanity. If something even remotely resembling these tragic adventures were to occur, if the thought we have seen constantly enlarge its domain over the centuries were to weaken, who knows whether this fact alone would not be sufficient to put an end to this period of mathematization, which seems, in antiquity as in modern times, to coincide exactly with the periods of true human progress?

292. *The Role of the Epistemologist*

But the role of the epistemologist is not to speculate on these eventualities, however near and threatening they may seem in some respects. It consists, it seems to us, in following the development of science, certainly not in order to teach science its business, but to carry out a task essentially different from that of the scientist himself, by trying to identify the thought processes he uses. This is an arduous task, for reason must succeed in separating the nature of its own forms from the content to which they are applied, a content without which they could not exist. In our opinion, however, this can be accomplished, at least partially, by means of a comparison with past scientific or extrascientific concepts, a comparison that in some sense allows us to isolate their common denominator. We should be gratified if the reader found our work even partially successful in achieving this goal.

NOTES

[1] See Ch. 13, n. 1.

[2] PRB 60; Eng. 42–43 [the bracketed insertion is Meyerson's].

[3] In a recent work Einstein gives the following justification of the exceptional status the theory of relativity attributes to the speed of light: "In order to give physical significance to the concept of time, processes of some kind are required which enable relations to be established between different places. It is immaterial what kind of processes one chooses for such a definition of time. It is advantageous, however, for the theory, to choose only those processes concerning which we know something certain. This holds for the propagation of light *in vacuo* in a higher degree than for any other process which could be considered, thanks to the investigations of Maxwell and H. A. Lorentz" (VVR 19; Eng. 28–29).

[4] Einstein, *Sitzungsberichte der Königlich preussischen Akademie der Wissenschaften, Physikalisch-mathematische Klasse*, 1923, pp. 32, 76, 137; Elie Cartan, in *Comptes rendus hebdomadaires des séances de l'Académie des Sciences*, 174 (1922): 437, 593, 734, 857, 1104.

[5] Planck, PRB 140 ff.; Eng. 96 ff. Cf. how the continuity of the static and dynamic fields, which is the great triumph of Maxwell's theory, would also be sacrificed in this case (PRB 165; Eng. 114).

[6] The proceedings of this *Congrès* have not yet been published as of this writing.

[7] ET 124, 127; Eng. 106, 108–109 [the bracketed insertions are Meyerson's].

[8] Cf. Edouard Branly's declarations in *Science et Voyages*, no. 16 (18 Dec. 1919).

APPENDIX 1. REVIEW BY ALBERT EINSTEIN*

Émile Meyerson: *La Déduction Relativiste* (Paris: Payot, 1925, 386 pages).

It is easy to say what is so unique about this book. It was written by a man who has grasped the pathways of thought of modern physics and who has penetrated deep into the history of philosophy and the exact sciences with a sure eye for psychological motives and interrelationships. Logical acumen, psychological instinct, multi-faceted knowledge, and straightforward expression are happily united here.

Meyerson's fundamental guiding principle seems to me to be that theory of knowledge cannot be achieved through the analysis of mind and through logical speculation but rather only by considering and intuitively grasping the empirical material. 'Empirical material' is in this case the actually given totality of scientific results and the history of their development. The author seems to have envisioned the following major problem: What is the relation of scientific knowledge to the essential complex of facts of experience? To what extent can one speak of an inductive method in science, to what extent of a deductive method?

Pure positivism and pragmatism are rejected, indeed ardently combatted. Subjective experiences [*Erlebnisse*] or facts of experience [*Erfahrungstatsachen*] are indeed the basis of every science, but they do not make up its content, its essence; rather they are merely the given to which science refers. Merely ascertaining empirical relationships between experimental facts cannot, according to the author, be represented as the only goal of science. First of all, such general relationships as are expressed in our laws of nature are not at all the mere ascertaining of the experiential; they can be formulated and derived only on the basis of a conceptual construction which cannot be extracted for us from experience as such. Secondly, science by no means contents itself with formulating laws of experience. It seeks, on the contrary, to build up a logical system, based on as few premises as possible, which contains all laws of nature as logical consequences. This system, or rather the structures occurring in this system, is coordinated with [*zugeordnet*] the objects of experience; reason seeks to arrange this system, which is supposed to correspond to the world of real things of prescientific Weltanschauung, in

252

such a way that it corresponds to the totality of facts of experience or (subjective) experiences. Thus, at the base of all natural science lies a philosophical realism. To trace all laws of experience back to what is logically deducible is the final goal of all scientific research according to Meyerson – a goal for which we always strive and which we nevertheless are vaguely convinced that we can achieve only incompletely.

In this sense Meyerson is a rationalist, not an empiricist. But he also differs from critical idealism in Kant's sense. For there is no feature, no characteristic, of the system we are seeking, about which we can know *a priori* that it must necessarily belong to this system due to the nature of our thought. This also holds for the forms of logic and for causality. We can only ask how the system of science (in its states of development thus far) is composed, but not how it *must* be composed. The logical foundations of the system as well as its structure are thus (from a logical point of view) conventional; their only justification lies in the performance of the system vis-à-vis the facts, in its unified character, and in the small number of its premises.

Meyerson sees in the theory of relativity a *new* deductive system of physics departing from previous physics; he denotes it – to designate its novelty formally – as 'relativism'. Here, he has gone too far in our opinion. The theory of relativity by no means claims to be a new system of physics. Starting from the view suggested by experiences with light, inertia, and gravitation, that there is no physically privileged state of motion (principle of relativity), it proposes the formal principle that the equations of physics ought to be covariant with regard to any arbitrary point transformations of the four dimensional space-time continuum. To this principle it adapts the basic laws of physics – as they were known before – with as few changes as possible. By itself the principle of relativity or, better, of covariance would be much too narrow a base on which to erect the edifice of theoretical physics. Thus, one cannot speak of a 'physics adapted to the principle of relativity' as a 'relativism', as a new system of physics. Inasmuch as Meyerson identifies the general claim of the 'principle of relativity', which asserts less, with the more far-reaching claim of 'relativism', he arrives, in my opinion, at a not quite accurate point of view with regard to the novelty of the theory and its claim to lasting significance. Not the theory as a whole, but only the adaptation to the principle of relativity, is new. [It seems to me that, everything considered, the author completely shares this point of view, for he often insists that relativistic thought is essentially in conformity with the laws and general tendencies science had already manifested earlier (*La déduction relativiste*, pp. xi, 61 and 227 ff., esp. 247 and 251 [pp. 4, 46 and 154 ff., esp. 165 and 167 above]).]

On the other hand, the principle of relativity, taken by itself, appears to be much better established by experience than is the *formal structure* of the theory as adapted to previous physical knowledge. We do not know yet, but we fear, that the concepts 'metrical field' and 'electromagnetic field' will turn out to be inadequate to the facts of quantum theory. But the view that thereby the principle of relativity as such could be toppled need scarcely be given serious consideration.

But we mention this only in passing. For Meyerson it was important that the system of thought of physics has acquired the character of a logically closed deductive system to a much higher degree than before by the adaptation to the principle of relativity. Meyerson does not fault this strongly deductive-constructive, highly abstract character of the theory, but rather finds that this character corresponds to the tendency of the whole development of exact science, which more and more sacrifices convenience of foundations and methods (in the psychological sense) to achieve a unified character of the system as a whole in the logical sense.

This deductive-constructive character prompts Meyerson to compare the theory of relativity in a very ingenious manner with Hegel's and Descartes' systems. The success of all three theories with their contemporaries is traced back to their logically closed, deductive character: the human mind wants not only to propose relationships, it wants to *comprehend*. The superiority of the theory of relativity over the other two theories is seen by Meyerson in the quantitative rigor of construction and in the adaptation to many facts of experience. Meyerson sees another essential character common to Descartes' theory of physical events and the theory of relativity, namely the reduction of all concepts of the theory to spatial or geometrical concepts; in relativity theory however, this is supposed to hold only after the inclusion of the electric [electromagnetic] field in the manner of Weyl's or Eddington's theory.

{I would like to deal more closely with this last point because I have quite a different opinion in this matter. I cannot, namely, admit that the assertion that the theory of relativity traces physics back to geometry has a clear meaning.} One can with better justification say that, with the theory of relativity, (metrical) geometry has lost its special existence vis-à-vis the instances of law-likeness which have always been denoted as physics [Meyerson is able to quote a passage from Eddington where he speaks of a 'geometrical theory' of the universe (*La déduction relativiste*, 137 [p. 96 above])]. Even before the proposal of the theory of relativity it was unjustified to consider geometry vis-à-vis physics as an '*a priori*' doctrine. This occurred only because it was usually forgotten that geometry is the study of the possible positions [and displacements] of rigid bodies. According to the general theory of relativity the metrical tensor determines the behavior of the measuring bodies and clocks as well as the motion of freely mobile bodies in the absence of electrical [electromagnetic] effects. The fact that this metrical tensor is denoted as 'geometrical' is simply connected to the fact that the appropriate formal structure first appeared in the science denoted as 'geometry'. But this by no means justifies denoting as 'geometry' every science in which this formal structure plays a role, even when in illustration for comparison one makes use of those notions which are customary in geometry. By similar reasoning Maxwell and Hertz could have denoted the electromagnetic equations of the vacuum as 'geometrical' because the geometrical concept of vector occurs in these equations. {It gives me satisfaction, furthermore, to emphasize that Meyerson himself in his last chapter explicitly says that the designation of the theory as 'geometrical' is actually without content; one could just as well say that the metrical tensor describes the 'state of the ether'.}

The essential point of the theories of Weyl and Eddington on the representation of the electromagnetic field lies not in the fact that they have incorporated the theory of this field into geometry, but that they have shown a possible way to represent gravitation and electromagnetism under a unified point of view, whereas beforehand these fields entered the theory as logically independent structures. [Consequently I believe that the term 'geometrical' used in this context is entirely devoid of meaning. Furthermore, the analogy Meyerson sets forth between relativistic physics and geometry is much more profound. Examining the revolution caused by the new theories from the philosophical point of view, he sees in it the manifestation of a tendency already indicated by previous scientific progress, but even more visible here — a tendency to reduce 'diversity' to its simplest expression, that is, to dissolve it into *space*. What Meyerson shows in the theory of relativity itself is that this complete reduction, which was the dream of Descartes, is in reality impossible.]

Furthermore, Meyerson correctly stresses that many presentations of the theory of relativity incorrectly speak of a '*spatialisation du temps*'. Space and time are indeed fused into a unified *continuum*, but this continuum is not isotropic. Indeed, the character of spatial contiguity remains distinguished from that of temporal contiguity by the sign in the formula giving the square of the interval between two contiguous world points. [The tendency he denounces, though often only latent in the mind of the physicist, is nonetheless real and profound, as is unequivocally shown by the extravagances of the popularizers, and even of many scientists, in their expositions of relativity.]

It is my conviction that Meyerson's book is one of the most valuable contributions on the theory of relativity which has been written from the viewpoint of epistemology. I only regret that the works of M. Schlick [and H. Reichenbach] seem to have escaped Mr. Meyerson, for he certainly would have been able to appreciate them.

NOTE

* This text is based upon the unpublished German text from the Einstein archive and upon the published French translation by André Metz which appeared as 'A propos de *La Déduction relativiste* de M. Émile Meyerson' in *Revue philosophique de la France et de l'étranger* 105 (1928) 161–166. Some passages which appear only in the German text are enclosed in curly brackets; others which appear only in the French text are enclosed in square brackets. At the suggestion of the Estate of Albert Einstein, we have preferred the German text where differences are evident; but that text may perhaps be the author's first draft. The changes may be those of Einstein, or they may have been modifications by Metz in consultation with Einstein. Einstein's role is evident, e.g., in the addition of the name of Hans Reichenbach to that of Moritz Schlick in the published version. (But we may note that Meyerson cites both, e.g., pp. 147, 153.) No later German text has come to our notice; the draft is in Einstein's hand. The English translation is primarily that of Peter Mclaughlin from the German text, edited by R. S. C., using David and Mary-Alice Sipfle's translation for material added in the French version.

APPENDIX 2. EINSTEIN–MEYERSON EXCHANGE, 6 APRIL 1922*

Meyerson: I should like to ask Mr. Einstein to clarify two points in particular which have less to do with the foundation of his conceptions than with the way in which they are often presented and with the conclusions people seem to want to draw from them. In the first place, we often hear the four-dimensional universe spoken of in terms suggesting that these four dimensions are all of an analogous nature. Now of course that is not at all accurate. To see this, one need only observe that since space is not infinite for Einstein, a person who continues to move in a straight line must eventually return to the same place – after a very long time, to be sure (a billion years, we are told, at the maximum velocity, which is that of light). It is obvious then that if the same were true of time, we ought, at some sufficiently remote future time, to return to the present moment. This would be tantamount to the very ancient concept of the Great Year introduced by the early Greek thinkers, particularly Heraclitus. The concept is found again in modern philosophers and scientists such as Nietzsche and Arrhenius, but here it would become as comprehensive and unexceptionable as it was for the ancients. That is not at all the opinion of Mr. Einstein, for whom the universe is, as you know, 'cylindrical', which is to say that it involves a curvature for the three spatial dimensions, from which curvature the fourth dimension, that of time, is exempt. But not only is it impossible for time to return cyclically, even in the very long term, as Einstein allows in the case of space; it is, as we all know, impossible for time to return or go backwards at all. We move in time altogether differently from the way we move in space. Granted there has been something of a change in this regard since the advent of the principle of relativity. Time no longer flows uniformly for everyone, and if a traveler returned from a trip carried out at a speed approaching that of light, his watch would not agree with those of the people who had remained in the same place. But there would be a limit to the divergence, for the traveler could never go backward in time. "One cannot telegraph into the past," Einstein quite rightly tells us, and out of the great upheaval which the theory of relativity necessitated for those concepts we believed the most firmly established, the principle of entropy is one of the two great principles of earlier physics to remain standing, the other being, as you know, the principle of least action.

257

The true situation in this respect seems to me to be as follows. Einsteinian mechanics implies reversibility. But this feature is not peculiar to Einsteinian mechanics; on the contrary, in classical mechanics phenomena appear equally reversible. Moreover, in both cases this is obviously a consequence of our deep-seated tendency to *spatialize* time, a tendency expressed here by the simple observation that in order to represent time we use a numerical duration – for any number is capable of being diminished as well as augmented.

As you know, in the domain of classical mechanics irreversibility is obtained by means of statistical considerations. These can obviously be retained in Einsteinian mechanics. Then, too, as has been suggested, perhaps it will be possible to combine the hypothesis of *quanta* with the principle of relativity. In any case, to avoid all ambiguity regarding this particular structural characteristic of 'dimension', it seems appropriate to speak of a universe not of four dimensions, but rather of $(3 + 1)$ dimensions, as Weyl has done. One must, however, bear in mind that this difference reflects not only the fact that in the formula for the interval the time variable is preceded by a sign different from those of the spatial variables, but reflects also, and above all, the fact of irreversibility. Our illustrious guest himself has enjoined us to consider the physical realities behind the symbols. Now this is a reality of the highest order, for it will not be any more possible in Einstein's world than it was in Newton's for us to walk into the past or to digest before we have eaten.

The second question is somewhat more complex. Relativity theory is generally represented as being the fulfillment, the concrete realization, one might say, of the program outlined by Mach. That is quite accurate in some respects, for insofar as the perfect relativity of motion in space is concerned, Mach was indeed one of your authentic forerunners. You will excuse me if I briefly remind my colleagues of the *Société de philosophie* what is at issue here. Everyone is acquainted with Newton's rotating bucket experiment: At the beginning, that is, when the sides of the bucket are already turning at a great speed but have not yet communicated their motion to the water, the water's surface remains smooth; then the liquid, as it too begins to rotate, climbs the sides of the bucket. Thus, concluded Newton, this rotational motion is an absolute motion, that is, unlike motion caused by attraction and depending on masses, it depends on the intrinsic essence of space itself. That is what Mach has contested. For him the motion of rotation depends equally on the masses present in space. If the centrifugal motion of the water appears to be independent of the rotation of the sides, it is because their mass is relatively insignificant. "No one could say what would happen if the sides became more and more massive until they reached, for example, a thickness of several kilometers."[1] It is on this point that Einstein's theory has sharpened Mach's concept. Indeed, for Einstein, gravitation and inertial motion, instead of being entirely separate, as they were for Newton, are, on the contrary, intimately connected, and we are now able to calculate the masses that need to be moved around a body in order to produce evidence of a centrifugal force perceptible to our measuring instruments.

But this aspect of Mach's theory, interesting as it is, is not its principal aspect, and the reason some people attempt to characterize the whole body of his thought as relativistic is that they are considering something quite different from the relativity of space. Mach is, in fact, above all a continuator of Auguste Comte. For him, as for the founder of positivism, science is only a collection of rules, of laws. All it knows and all it need seek to clarify are relationships, connections between things; it must resolutely set aside anything aiming at knowledge of things themselves, a knowledge that is declared to be metaphysical. You know, moreover, that positivism, from the time of the earliest disciples of Auguste Comte, has often tried its best to align itself with an extreme idealism (as, for example, Taine has done with the doctrine of Hegel), going from the preclusion of any search for the nature of the things to the claim that things do not exist outside our consciousness. Therefore, one should not be surprised, especially in the light of the confusion encouraged by the use of the somewhat ambiguous term *relativity*, to see the proponents of this doctrine attempt to appeal to the authority of Einstein's ideas by proclaiming that the relativity of space proves the relativity of our knowledge at every level, and consequently shows how vain it would be to attempt to get to the very core of things, as atomic theories aspire to do. That is actually the crucial point of the entire question. Comte, driven by his powerful scientific instinct, had, with a felicitous bit of illogic, declared atomism a 'good theory.' John Stuart Mill, however, had already noted that if we are to follow the principles of positivism more rigorously, we must disregard the object and seek to establish direct relations among our sensations, and Mach showed himself to be resolutely hostile to the atomic theories. For him, as for the energeticist school which followed him, the ideal of a science is exemplified by thermodynamics, because it seems to renounce any representation of the matter with which it is concerned and limits itself to deducing its statements from two abstract principles. Such an attitude manifested itself quite markedly in the last years of the nineteenth century and the beginning of the present century at the time of the discoveries that revealed the discontinuity of matter and thus inspired a return to atomistic concepts. In the eyes of the energeticists, this powerful evolution, which clearly constituted an immense progress for knowledge, appeared to be a disastrous step backwards.

I shall not attempt here to demonstrate to you how vain these pretensions were — pretensions which science has resolutely ignored. Suffice it to say that on this point there seems to be no really close connection between Mach's conceptions and Einstein's theory. One can quite easily support the relativity of space and nevertheless be convinced, as Malebranche had already established, that no science is possible unless one first posits an object situated outside consciousness, and that consequently science cannot dispense with specifying how it conceives of the object, through the modifications which the progress of our knowledge imposes on our picture of it. It seems to me, moreover, that Einstein's own attitude confirms this way of looking at things. In fact, the evolution toward atomism of which I spoke a moment ago, and which made dyed-in-the-wool energeticists so unhappy, owes much to Einstein himself. In 1905, at almost the same time as his first works on relativity, he published a work in which, without knowledge of Gouy's results in particular or of Brownian motion in general, he determined the amplitude of molecular motion, and, as we know, his formulae were subsequently used by Perrin. Furthermore, when the question of the phenomena of black body radiation, so strange and puzzling for atomic physics, came up at the 1911 *Conseil de physique*, a purely phenomenalistic approach was explicitly suggested to the participants; Einstein, however, insisted quite clearly and vigorously on the necessity of presenting what we know of these phenomena 'in concrete form', that is, by a representation in space and by means of a mechanism truly able to explain the observations. He emphasized, moreover, that such an image should be as complete and coherent as possible, pointing out the difficulties and improbabilities one comes up against here if one adopts Planck's hypothesis. I do not think I am going too far in supposing that Mr. Einstein himself is far from sharing Mach's opinion in this area. But I believe that given the particular interest this question holds for us, not only from the point of view of the theory of our scientific knowledge — of epistemology — but also from the standpoint of philosophy in general, some clarification by the author of the theory of relativity himself would be very useful.

Einstein: In the four-dimensional continuum, it is certain that all directions are not equivalent.

On the other hand, there does not seem to be any close relationship, from the logical point of view, between the theory of relativity and Mach's theory. Mach draws a distinction: on the one hand, there are things, which we cannot change — these are the immediate data of experience — and, on the other hand, there are concepts, which we can modify. Mach's system studies the relationships that exist among the data of experience; for Mach the ensemble of these relationships is science. This is an unsatisfactory way of looking at things; essentially what Mach has done is make a catalog, and not a system. As good as Mach was as a physicist, he was every bit as deplorable as a philosopher. His narrow view of science led him to reject the existence of atoms. It is likely that if he were still alive today he would change his mind. I must, however, agree with Mach on one point: concepts can change.

NOTES

* *Bulletin de la Société française de Philosophie* 17 (1922) 107–112; from the proceedings of the meeting held 6 April 1922.
[1] [Cf. Ernst Mach, *The Science of Mechanics*, trans. Thomas J. McCormack (Chicago: Open Court, 1907) p. 232.]

NAME INDEX